MAMMALS IN WILTSHIRE

WILTSHIRE ARCHAEOLOGICAL AND NATURAL HISTORY SOCIETY

Marion Browne recording bats, early 1980s

MAMMALS IN WILTSHIRE

A TRIBUTE TO THE LIFE AND WORK OF MARION BROWNE

(21.4.21 – 16.7.90)

Patrick Dillon

in conjunction with

Sally Scott-White and Paul Jones

Wiltshire Biological Records Centre

Wiltshire Archaeological and Natural History Society

Published by
Wiltshire Archaeological and Natural History Society
41 Long Street, Devizes SN10 1NS
1997

ISBN 0 947723 05 6

A catalogue entry for this book is available from the British Library

Cover photos by Marion Browne

Origination by Avonset, 1 Palace Yard Mews, Bath BA1 2NH

Printed and bound by Antony Rowe Ltd, Bumper's Farm, Chippenham, Wiltshire

FOREWORD

Numerous books have been written on different aspects of mammals in Britain. *Mammals in Wiltshire* represents a first attempt to summarise and present a systematic picture of the situation in this county. For the contents of this book we are indebted to the work and enthusiasm of a remarkable naturalist, Marion Browne, who from the 1970s until her death in 1990 built up a wide range of friends and associates who provided her with systematic annual records of their observations on mammals. These records, in conjunction with the information accumulated by Marion herself, provide the materials for this book. Marion Browne was a person of great kindness and generosity and on her death she donated a considerable sum of money to the Wiltshire Archaeological and Natural History Society for the furtherance of her work on natural history in the county. It is this legacy which has made possible the publication of *Mammals in Wiltshire* in her memory.

Patrick Dillon, the author of the book and a long-term friend, collaborated with Marion Browne in her mammal researches. As long ago as 1976 they evolved a scheme for collecting systematic data on mammals with the intention of publishing a definitive work. Patrick himself is also an accomplished naturalist with a particular interest in the ecological history of mammals. In writing the book, he has been able to incorporate his own historical information with Marion Browne's data and so produce a comprehensive record. In re-working Marion's data he has been assisted by a colleague Paul Jones who is also a zoologist.

Of particular interest to students of mammals is the frequency of their occurrence and the distribution of each species. Here we owe a great deal to Sally Scott-White of the Wiltshire Biological Records Centre for the inclusion in her database of the records from Marion Browne's survey and from them the construction of distribution maps which provide an important part of the book.

By way of conclusion, how can we sum up the achievements of Marion Browne and her associates? One thing is certain, their unique observations and records have opened up a new era in the study of mammals in Wiltshire. More than that, by making so much useful information available in a readily understandable form, I believe they have provided many valuable opportunities that previously did not exist to students at all levels. Let us hope that this book may be a stimulus for an increasing study on mammals in schools and all organisations associated with natural history.

Wilfrid Dowdeswell

CONTENTS

PREFACE

This account of mammals in Wiltshire is dedicated to the life and work of Marion Browne.

Marion Browne was born at Warsash, Hampshire on 21 April 1921 the daughter of Patrick and Sylvia Hall. Although there was a strong natural history influence in the family – Patrick Hall was a botanist with a particular interest in orchids and Kate Hall, a great aunt, was Curator of the Borough of Stepney Museums and author of *Nature Rambles in London* (Hodder and Stoughton 1908) and *Common British Animals* (Bartholomew Press 1913) – Marion's early interest was in handicrafts.

Marion Browne attended Rookesbury Park Preparatory School, Wickham, Hampshire and then boarded at Prior's Field, Godalming, Surrey before going to the Paris Academy of Dressmaking, London. The War interrupted her studies and she served in the Women's Land Army for its duration. After the War she returned to the Paris Academy and completed a Diploma before joining Motley Theatrical Costumiers as a costume maker. Later she became Needlework Editor of *Good Housekeeping* and wrote two books on needlecraft. In 1953 she married Ronnie Browne, Managing Director and later Chairman of T.B. Browne Ltd, a London based advertising agency.

Marion Browne moved to Wiltshire with her husband in 1966. They lived at Latimer Lodge, West Kington. Marion Browne's work in natural history in Wiltshire was through both the Wiltshire Trust for Nature Conservation and the Wiltshire Archaeological and Natural History Society (WANHS). From 1973, until the appointment of a full-time development officer in 1979, she organised the 'sales stall' at Trust events, and from 1973 until 1988 she served on its Council. She was a member of the Committee of the Natural History Section of the Society from 1974, and Natural History Editor of the *Wiltshire Archaeological and Natural History Magazine (WAM)* from 1980 until her death on 16 July 1990.

Although a naturalist in the general sense, Marion Browne's particular interest was in mammals and, to a lesser extent, amphibians and reptiles. She combined practical work in conservation with the care of injured and orphaned animals, extensive field studies, and biological recording. Her work was recorded in meticulous detail in working files and diaries supplemented with her own line drawings and photographs. This book is concerned with Marion Browne's contribution to biological recording in Wiltshire and constitutes a systematic account of the mammals of the county. It brings to fruition a project which she started in the mid-1970s.

Before the 1970s, little information had been published on the status and distribution of mammals in Wiltshire. Checklists, with impressionistic statements about status, were published by Hony in 1917 and Peirson in 1923. Occasional notes and records were published in *WAM*, the *Marlborough College Natural History Report*, the *Annual Report of the Salisbury and District Natural History Society*, and the *Dauntsey Flora and Fauna List.* In the 1960s, *WAM* carried three substantial papers on mammals: a systematic account of the distribution of badgers in the county; a local survey of road casualties; and an account of the rabbit warren on Fyfield Down.

During the 1960s, Beatrice Gillam, then County Mammal Recorder, and Noel King of the (then) Nature Conservancy, explored the possibility of compiling a book on mammals in Wiltshire. Noel King moved to Herefordshire in the early 1970s before any substantial data had been collected although records for the period 1970-1972 were published in *WAM* in 1973. At about the same time Marion Browne became involved in natural history locally through fieldwork, first on the fox, and then on the harvest mouse. Her early observations on foxes were published in *WAM* under the pseudonym 'Renarde'. In 1974 she published a note jointly with Beatrice Gillam on the harvest mouse in Wiltshire and in 1975 became Recorder of Amphibians, Reptiles and Mammals.

It became apparent that a definitive publication on mammals in Wiltshire could be prepared only after a systematic survey on the status and distribution of species had been carried out. After some discussion by the Committee of the Natural History Section of WANHS, it was agreed that Marion Browne should coordinate the survey and that I should edit the ensuing publication. It was envisaged that the publication would consist of two parts: (i) a history of mammals in Wiltshire and (ii) the results of the systematic survey. I was to take responsibility for the first part and Marion Browne for the second.

The survey began in 1976. Despite an encouraging response, it soon became clear that it would take some years to collect the data necessary for a definitive publication and so an alternative publishing strategy was agreed. The collated records were published annually from 1977 to 1988 (the survey effectively ended in 1987) as a series of 'Mammal Reports', first in *WAM*, then in the *WANHS Annual Report* and finally in *WANHS Natural History County Recorders' Reports*. By the early 1980s, Marion Browne had gathered enough data for a series of detailed studies on the status and distribution of individual species or groups of species. These were published each year in *WAM* until Marion Browne's death in 1990 and covered the voles, mice, dormouse, shrews, hedgehog, mole, and common rat.

This book presents, for the first time, a comprehensive account of mammals in Wiltshire. It brings together the full data from the survey conducted by Marion Browne and the information which I collected on the history of mammals in Wiltshire. For each species, there is a summary statement on status, a map showing known distribution, and details of the number and nature of the records collected. For some species, information is provided on habitats, field signs, mortality and seasonal activity. There are a number of line drawings, all produced by Marion Browne to illustrate her published papers and diaries. The book ends with appendices which list (i) the people who supplied Marion Browne with information for the survey; (ii) works written by or about Marion Browne; and (iii) Marion Browne's works in manuscript.

The publication of this book has been made possible through Marion Browne's legacy to WANHS. I would like to thank the Natural History Committee of the Society for inviting me to prepare the book and for offering advice, first under the chairmanship of Michael Fuller and then Phillida Sneyd. Jennifer Harrison and Elizabeth Gale of the (then) Wiltshire Trust for Nature Conservation answered my queries about Marion Browne's work with the Trust. Christopher Perraton commented on the text and Pat Morris on early versions of the species accounts. Mick Peacey read the entire manuscript and checked the technical detail. Christopher Bindon, Jean Rayment and Philippa Noad helped me with the historical survey during periods when they were employed at Devizes Museum between 1976 and 1981.

I owe a particular debt to Sally Scott-White of the Wiltshire Biological Records Centre, which operates under the joint auspices of the Society and Wiltshire County Council, and Paul Jones, a freelance researcher. Sally Scott-White arranged for the information from Marion Browne's survey to be entered in the database at the Wiltshire Biological Records Centre and provided me with distribution maps and summary data. Paul Jones used the summary data to produce the pie charts and histograms which accompany the species accounts. Kate Fielden, the Society's Editor, offered welcome advice about the preparation of the typescript, and staff of the Biological Records Centre guided the work through the final stages of production. Finally, and most important, I would like to thank the many people who supplied Marion Browne with records and corresponded with her during the eleven years of the survey. A list of these people, compiled from her card index and the credits in her *Annual Reports*, appears as an appendix; any omissions are my responsibility alone. I hope this book represents a fitting testimony to the enthusiasm of these people as well as a tribute to the work of Marion Browne.

Patrick Dillon
Kingston Winslow, 1996

ABBREVIATIONS

The following abbreviations have been used:

BBONT	Berkshire, Buckinghamshire and Oxfordshire Naturalists' Trust
DFFL	*Dauntsey Fauna and Flora List*
MCNHS	Marlborough College Natural History Society
SDNHS	Salisbury and District Natural History Society
WAM	*Wiltshire Archaeological and Natural History Magazine*
WANHS	Wiltshire Archaeological and Natural History Society
WBRC	Wiltshire Biological Records Centre
WTNC	Wiltshire Trust for Nature Conservation (now the Wiltshire Wildlife Trust)
WNQ	*Wiltshire Notes and Queries*

INTRODUCTION

The accounts of mammals in Wiltshire which follow are based on pre-1976 records extracted from published and manuscript sources, and information collected through a survey which ran from 1976 to 1987 inclusive.

Pre-1976 information

For each species a provisional distribution map, as of 1976, was established from records extracted from the national Biological Records Centre and from known local sources (listed in Dillon and Noad 1980 and Dillon 1984). These 'historical' records were entered onto index cards and made available to Marion Browne on a species by species basis as she wrote her systematic accounts for *WAM*. For the rats, the historical data were thought to be of sufficient interest to justify a separate article in addition to the systematic account (Dillon 1990).

As the *WAM* accounts covered just a few species, a large proportion of the historical data were not used. This publication provides the first opportunity to publish the significant historical data. 'Significant historical data' are those that indicate a substantial change in the status of a species. For some species these data chart decline and ultimate extinction; for others they chart introduction and subsequent colonisation.

The use of the historical data does present some anomalies: for example, the wild boar and the wolf are now extinct in Britain but there are early records of both for Wiltshire. The wild cat, although not extinct, is considerably restricted in range; it almost certainly existed in Wiltshire although there are no records of it. For the sake of completeness, all three species have been included. Leisler's bat, which is very difficult to distinguish from its larger relative the noctule, has been recorded in central England but not in Wiltshire; it has not been included. The beaver has not been included because, although it may have survived in England into historical times, there are no references to it in Wiltshire. The mouse-eared bat, thought to be extinct in Britain, has not been recorded in Wiltshire and is not included.

The card index of historical data has been placed in the Wiltshire Biological Records Centre. The index also includes some records, not used in this publication, of fossil and sub-fossil assemblages and mammals recorded as part of the stratigraphy during excavations of prehistoric sites.

The 1976-1987 Survey

In 1976 a record sheet was designed and sent out to a (cumulative) list of correspondents and distributed through local natural history societies, conservation organisations, local organisations such as the Women's Institute, and to anglers, water bailiffs, farmers and estate workers. Illustrated lectures were given and workshops were held in support of the survey and occasional articles were published in the local press.

Respondents were asked to provide the following information about their observations:

- locality and grid reference;
- date;
- species;
- nature of record (live, dead – on road or elsewhere – or field signs such as droppings, burrows, footprints, and skeletal material); and
- habitat.

If in doubt about any of the information, respondents were invited to telephone Marion Browne or send material such as droppings, bones, and nuts opened by small mammals, for identification.

In addition, questions were asked of people during conversations from which transcripts were made and records extracted.

Occasionally, field meetings were held to survey a specific area for evidence of mammals in general or to survey a wider area for evidence of a particular species.

All responses were acknowledged by Marion Browne in writing. Over the years, she built up an extensive network of correspondents (appendix 1), many of whom submitted records regularly. The collated records were published annually from 1977 to 1988 as a series of 'Mammal Reports', first in *WAM*, then in the *WANHS Annual Report* and finally in *WANHS Natural History County Recorders' Reports* (Browne 1977-1988). A series of detailed studies on the status and distribution of individual or groups of species were published in *WAM* (Browne 1983-1990; Dillon and Browne 1984). An earlier paper on the harvest mouse (Dillon and Browne 1975) may be considered part of this series as it was based on a systematic survey conducted under the auspices of the Mammal Survey; the data were subsequently subsumed into the Wiltshire survey.

Record sheets and associated notes and correspondence are now held in the Wiltshire Biological Records Centre. Summary data, in the format given above, along with a note on the existence of any supporting correspondence, have been entered in the database at the Centre.

Format for the species accounts in this book

The species accounts are in the following format:

Nomenclature

Order; Family; English name; scientific name.

Description

Brief notes on the appearance of the species and characteristics used in identification; head and body (excluding tail) length.

Standard and summary publications

Pagination is provided for the relevant pages of standard works – *The Handbook of British Mammals* (Corbet and Harris 1991), the *Atlas of British Mammals* (Arnold 1993) and *A Review of British Mammals* (Harris, Morris, Wray and Yalden 1995).

For each species, the *Handbook* includes summary information on recognition, measurements, variation, distribution, history, habitat, social organisation and behaviour, feeding, breeding, population, predators and mortality, parasites and diseases, and relations with Man. The *Atlas* includes distribution maps and tabulated summary data on status. The *Review* includes summary information on status, distribution, population, including a numerical estimate for each species, historical changes and population trends and threats.

For those species, or groups of species, which are threatened and/or are afforded legal protection, pagination is given for the relevant sections of *A Red Data Book for British Mammals* (Morris 1993a). This book includes summary information on legal status, distribution, population size, history and perceived threats, an ecological and research overview, and priorities for future action. The last section gives information on conservation-related research and its application and practical conservation measures.

For those species which are deemed to have economic significance as game animals and their predators, pagination is given for the relevant sections in *Game Heritage: An Ecological Review from Shooting and Gamekeeping Records* (Tapper 1992). This book includes summary information, in atlas, tabulated and graphical formats, of changing status based on the number of animals killed.

These five publications are referred to respectively as *Handbook, Atlas, Review, Red Data Book* and *Game Heritage* .

Status

This is a two part statement on status nationally and is based on the entries in the *Atlas of British Mammals* (Arnold 1993), *A Red Data Book for British Mammals* (Morris 1993a) and *A Review of British Mammals* (Harris, Morris, Wray and Yalden 1995).

The first part indicates whether the species is native or introduced and gives a generalised statement of abundance. Terms such as 'rare', 'declining', and 'vulnerable' are used. It is recognised that, in most cases, these terms have no absolute meaning; they are used in general and relative ways. The 'common' status attached to some of the species which are not thought to be at risk, and for which no detailed survey has been undertaken, may be questionable (Morris 1993b).

The second part is based on *A Review of British Mammals* (Harris, Morris, Wray and Yalden 1995) and, for each species, gives a numerical estimate of the pre-breeding population of Britain and any pertinent information on historical and projected population trends and threats. As the authors of the *Review* acknowledge, their numerical estimates are bound to be controversial but they hope to provide a baseline from which future changes in status may be measured. In the *Review*, the authors provide a detailed explanation of how they arrived at the estimates including reliability scores which give some indication of the level of accuracy thought to have been achieved.

National context

This is a generalised, summary statement of distribution and habitat preferences based on the *Handbook of British Mammals* (Corbet and Harris 1991) and the *Atlas of British Mammals* (Arnold 1993).

The Wiltshire accounts

For each species, these open with summary statements about any special surveys, appeals for records and references to previously published accounts in the county. This is followed by information on:

Number and nature of records

The total number of records held on the database at the Wiltshire Biological Records Centre is given. Some of these are imprecise and so a second figure, of records used in the analysis, is given. When sufficient data are available, they are presented as a pie chart, showing percentage representation and numbers of records, under the categories 'live', 'dead', and 'field'. The last of these includes all field signs such as nests, burrows, runs, droppings, feeding signs, presence in bird pellets, etc.

The term 'record' relates to the occurrence of a given species in a given place, irrespective of the numbers of animals observed. Marion Browne, in her *WAM* publications, used 'record' to denote the number of animals of a given species in a given place, and so in all cases her figures are correspondingly higher.

Distribution

Records are shown by 1km square on an outline map of Wiltshire. In the interests of confidentiality, maps are omitted for the badger and otter. Records made since 1990 have also been included in the maps for completeness and are shown by open circles (o).

A distribution map is seldom a reflection of reality. In most cases it is little more than a record of where recording has been undertaken. It follows that all species are under-recorded. For some, the under-recording is particularly apparent and attention has been drawn to it. It is sometimes the practice to present distribution maps relative to maps which show the distribution of recording effort or where a species has been looked for but not found (so called 'negative' records). There are not sufficient data to justify either of these practices in this publication. However, fig. 0.1 is a composite distribution map, which shows all of the 1km squares for which there is a record of at least one species. It follows that the blank areas on this map are the parts of the county for which mammals are most under-recorded. These correspond particularly with the triangle formed by Swindon, Malmesbury and Cricklade (which includes Braydon Forest), the Vale of Pewsey, and the extreme south-west of the County with the exception of parts of the Vale of Wardour.

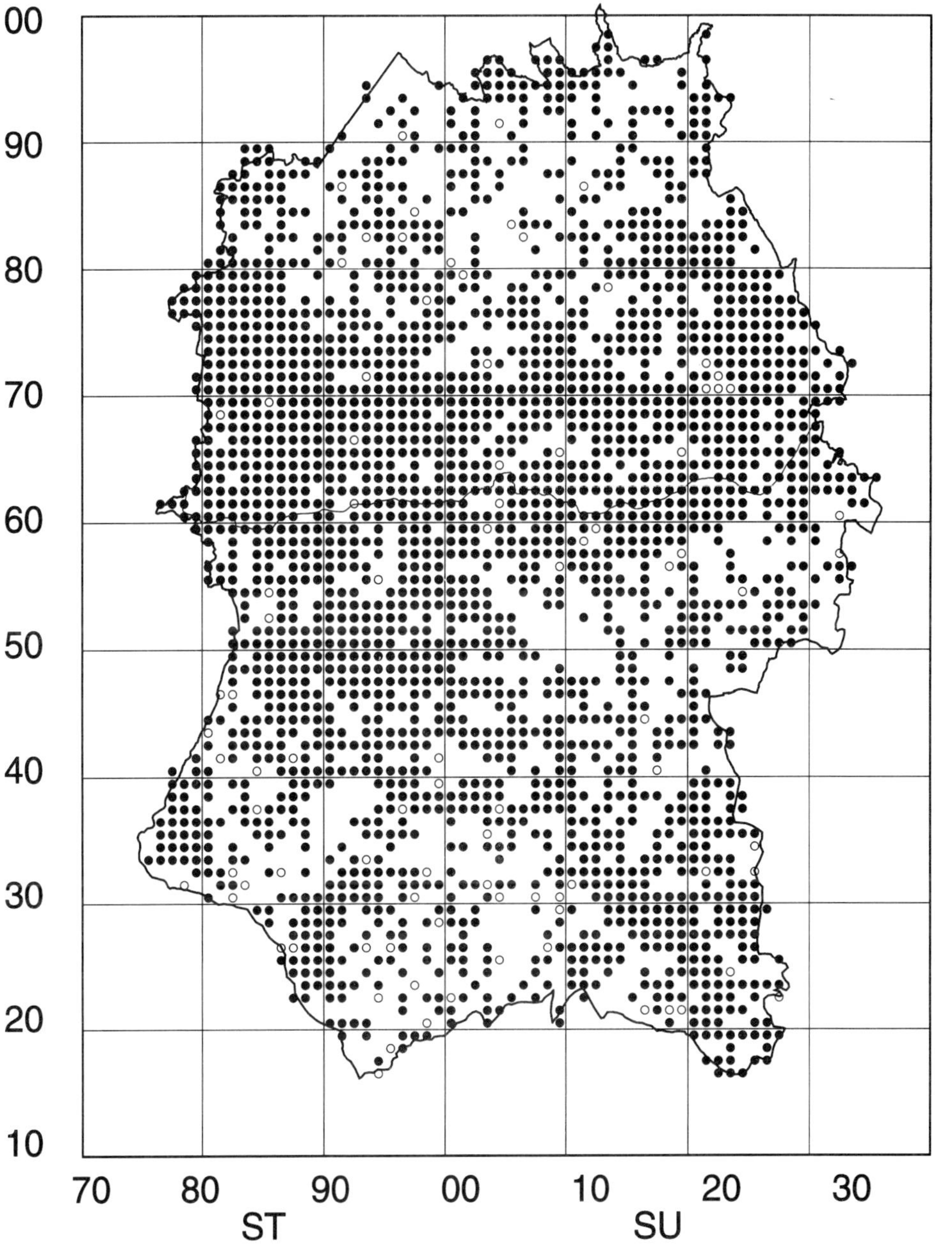

Fig. 0.1. Composite distribution map showing all of the 1km squares in Wiltshire for which there is at least one mammal record

Populations of most mammals are discontinuous for reasons to do with habitat selection, habitat availability, dispersal rates, population dynamics and vulnerability to extrinsic factors (Bright 1993). For most species, the reasons for discontinuities in distribution will not be evident from the maps; indeed the distribution data have not been subjected to the types of analysis that would demonstrate such relationships.

Habitats

The data are presented in a histogram, showing percentage representation and numbers of records, in specific habitat types grouped under the broad categories 'open', 'marginal', 'artificial' and 'closed'. These four categories are further sub-divided to reflect land-use (fig. 0.2).

- open
  - wasteland
  - downland
  - parkland
  - arable
  - pasture
  - wetland
    - water meadow
    - cress bed
    - river valley

- marginal
  - pond/lake
  - river
  - canal
  - dry wall
  - hedgerow
  - road verge
  - railway bank

- closed
  - woodland
  - plantation
  - coppice

- artificial
  - playing field
  - airfield
  - institutional grounds
  - churchyards
  - garden
  - outbuilding
  - house

Fig. 0.2. Habitat types used in the 1976-1987 survey of mammals in Wiltshire

Judged by the standards of the 1990s, this habitat classification, devised at the start of the survey in 1976, and sent to correspondents to be used in support of their recording, looks idiosyncratic. It was devised before the issue of a standardised habitat classification was addressed at the local level through the work of the Wiltshire Biological Records Centre, and at the national level through the work of the Institute for Terrestrial Ecology (ITE). At the time, it was thought to be a workable synthesis of what was known about the habitats of mammals generally and what was known specifically about the habitats available in Wiltshire. It is thus influenced heavily by Grose's *Flora of Wiltshire* (1957), the then definitive work.

Ideally, data for both habitat and distribution should be considered alongside the topography of the county, its geology and land-use. It has not been possible to conduct such an analysis for this publication but the *Wiltshire Flora* (Gillam 1993) provides a useful review of the habitats in the county relative to these factors. There is now a move to define the land areas of Britain in terms of geological and topographical characteristics and the types of habitat they support, so called 'natural areas', rather than by administrative boundaries so that more meaningful conservation policies may be devised. The map of England published by English Nature (*English Nature* No. 7 1993) defines 76 natural areas. For Wiltshire, running north to south, the major ones are now (1996) Cotswolds, Thames and Avon Vales, Berkshire and Marlborough Downs, Wessex Vales and South Wessex Downs. All of these natural areas run into neighbouring counties and may offer more meaningful units with which to equate mammal status at the regional level. At the habitat level, the 32 'land classes' developed at the ITE (see, for example, Bunce, *et al.* 1996) or variants of it (see, for example, Cresswell, Harris and Jefferies 1990) appear to offer potential for interpreting mammal distribution relative to landscape. The categories used in the survey reported here are broadly compatible with the ITE land classes.

Field Signs

Where sufficient data are available, these are presented in a histogram (percentage representation and numbers of records) showing a breakdown of the nature of the records – for example, nests, burrows, runs.

Mortality

Where sufficient data are available, these are presented in a histogram (percentage representation and numbers of records) either as a series of monthly totals or as a breakdown of the nature of the records – for example, dead on road, drowned, killed by domestic cat.

Seasonal activity

Where sufficient data are available, these are presented in a histogram (percentage representation and numbers of records) showing monthly totals.

Illustrations

For some species, line-drawings produced by Marion Browne illustrating the appearance of the animals and their skulls, jawbones and dentition and feeding and other field signs are included. Most of these were used as illustrations in her *WAM* articles; a few have been extracted from her notebooks and have not been published before.

References

Details of the printed sources of historical records and references to other literature are given in full in the references section. Some of the sources, for example, the *Bulletins* of the Natural History Section of WANHS and the Salisbury and District Natural History Society and some of the manuscript and privately printed materials are not available through public libraries. They may be viewed, by appointment, in the WANHS Library in Devizes.

An overview of the Wiltshire data

An overview of the status of mammals in Wiltshire is given in fig. 0.3 which shows the number and percentage of 1km squares in which each of the species occurs. Like the individual distribution maps, the list says as much about the ease with which different species can be recorded and the effort put into recording them as it does about their relative status. It does, however, highlight the under-recording of certain species, particularly the smaller mammals and some of the bats.

Of the ten most frequently recorded species in Wiltshire, seven are in the top ten in the national ranking list (Arnold 1993). The red squirrel, pipistrelle and otter are in the top ten in the national but not the Wiltshire list. Although the red squirrel and otter have a restricted distribution in Britain, a comparatively large amount of effort has been invested in recording them through special surveys. The red squirrel is no longer found in mainland central southern England, and the otter is at a very low population level in the region. The badger and brown hare are higher placed in the Wiltshire than in the national list. The county offers very good habitats for both species and, in the case of the badger, the population is monitored by a county badger group.

It follows that the frequency with which a species is recorded is not necessary a reflection of the relative abundance of that species. It is interesting to compare the data in fig. 0.3 with the estimated British pre-breeding population size of each of the species. Of the six species with estimated populations over 20,000,000 – field vole, common shrew, wood mouse, rabbit, mole and bank vole, only two, rabbit and mole, have been recorded in more than 10% of the 1km squares in Wiltshire. By contrast, the badger and brown hare, recorded in 30% and 28% respectively of 1km squares in Wiltshire, have estimated British populations of 250,000 and 817,500 respectively.

	No. 1km squares	% 1km squares
Badger	1033	30
Brown hare	990	28
Fox	828	24
Rabbit	730	21
Mole	692	20
Hedgehog	601	17
Roe deer	543	16
Grey squirrel	415	12
Weasel	325	9
Common rat	279	8
Pipistrelle	277	8
Water vole	241	7
Stoat	241	7
Brown long-eared bat	205	6
Common shrew	189	5
Wood mouse	168	5
Harvest mouse	158	5
Field vole	145	4
Bank vole	125	4
Pygmy shrew	89	3
Fallow deer	87	2
Water shrew	86	2
Serotine	73	2
House mouse	67	2
Mink	53	2
Common dormouse	51	1
Daubenton's bat	45	1
Natterer's bat	41	1
Muntjac	41	1
Noctule	37	1
Greater horseshoe bat	33	1
Lesser horseshoe bat	29	1
Whiskered bat	26	1
Otter	20	1
Yellow-necked mouse	16	<1
Red deer	11	<1
Feral ferret	11	<1
Brandt's bat	7	<1
Barbastelle	6	<1
Bechstein's bat	4	<1
Sika deer	4	<1
Coypu	3	<1
Red squirrel	2	<1
Ship rat	1	<1

Total number of 1km squares = 3500

Fig. 0.3. Frequency of occurrence of mammals in Wiltshire by 1km square

As Morris, in his introduction to *The Red Data Book for British Mammals* has pointed out, mammals are susceptible to major changes that take place too fast for normal adaptive processes to bring about the ecological adjustments necessary for survival (Morris 1993a). In a predominantly rural county like Wiltshire, agriculture and land-use will be the dominant influences for most species. For some, there may have been a decline since the end of the survey reported in this publication: concern has been expressed in recent years about the water vole and brown hare. Other species, for example, greater horseshoe bat, lesser horseshoe bat, common dormouse and red deer have genuinely restricted distributions in Wiltshire owing to the availability of suitable habitat or being at the edges of their range. Such populations may be susceptible to even small changes in agricultural practice or land-use.

Some species have legal protection through the Wildlife and Creatures and Wild Plants Act of 1975 and the Wildlife and Countryside Act of 1981 and their subsequent amendments. Of the species found in Wiltshire, the following have 'full' protection under schedule 5 of the Wildlife and Countryside Act: all species of bat, common dormouse and otter. The following have 'partial' protection under schedule 6: hedgehog, the shrews and the badger. Protection, whether full or partial, is seldom directed towards addressing the causes of decline.

A SYSTEMATIC ACCOUNT OF MAMMALS IN WILTSHIRE

Order Insectivora — Family Erinaceidae

Hedgehog — *Erinaceus europaeus*

Easily recognisable, spined, insectivore with rotund body. Head and body approximately 260 mm. Foot, toe pads (fig. 1.1), skull, jawbones and dentition (fig. 1.2) are diagnostic.

Handbook, 37-43; *Atlas*, 22-23; *Review*, 7-9; *Red Data Book*, 8-13; *Game Heritage*, 70-71.

Status

Native. Locally common.

The total pre-breeding population in Britain is estimated at 1,555,000. A steady population decline is suspected, probably attributable to mortality on roads, habitat changes and, in some regions, predation by badgers.

Nationally

Widespread distribution, particularly woodland edge, meadowland with hedgerows and urban and suburban areas with mature gardens.

Wiltshire

Special appeal for records made 1984; data up to 1985 published in *WAM* 81, 111-122.

1749 records, of which 1658 have been used in this analysis. 75% of the records are of dead animals (fig. 1.3). Widespread distribution, particularly in the north-west of the county and in river valleys; probably under-recorded on the chalklands (fig. 1.4). Most records are from roadside habitats and gardens (fig. 1.5). Mortality (fig. 1.6) and seasonal activity (fig. 1.7) show similar patterns, that is, most records April-November, with a peak in June; there are comparatively few records December-March.

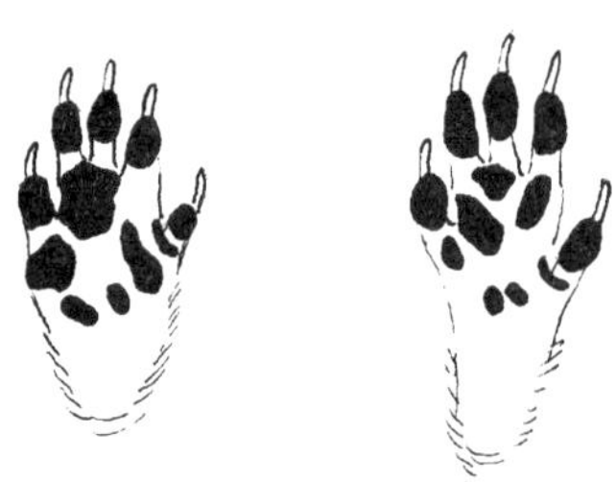

Fig. 1.1. Hedgehog: foot and toe pads, approximately x ⅔

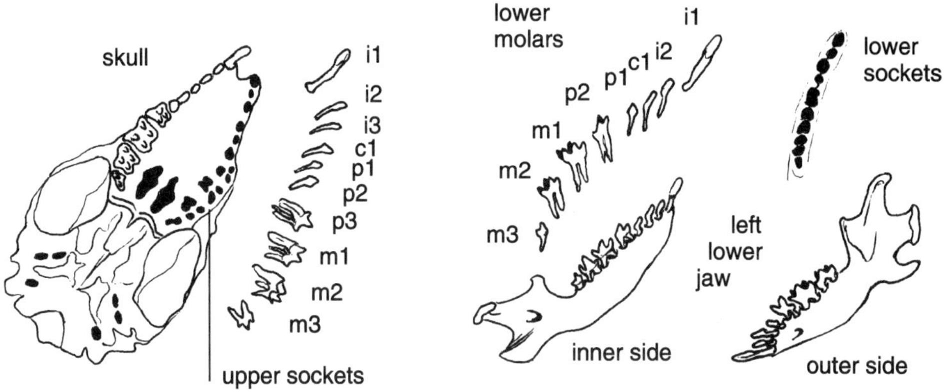

Fig. 1.2. Hedgehog: skull, jawbones and dentition, approximately x ⅔
i, incisor; c, canine; p, premolar; m, molar

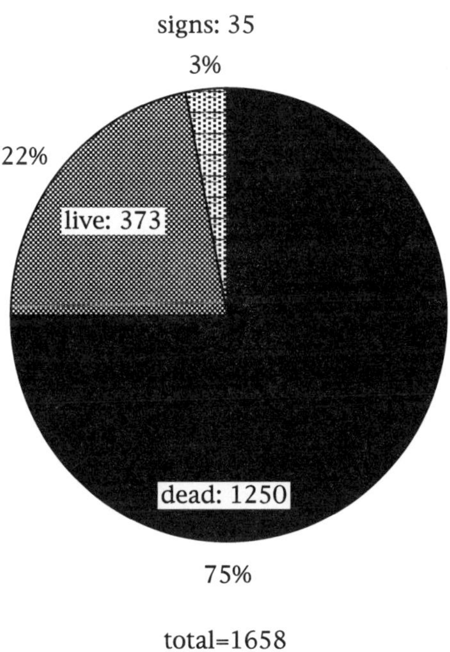

Fig. 1.3. Hedgehog: Number and nature of record

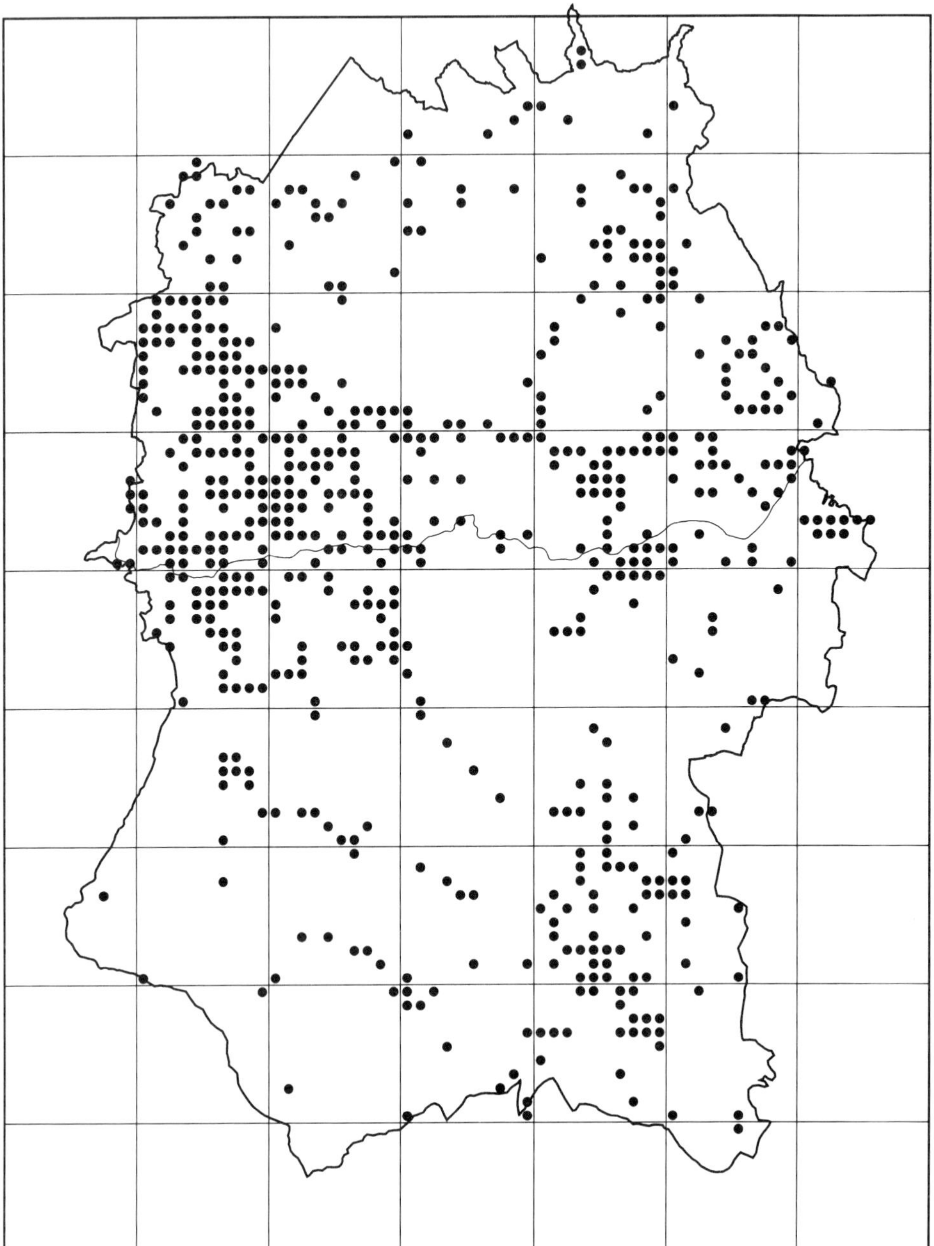

Fig. 1.4. Hedgehog: distribution

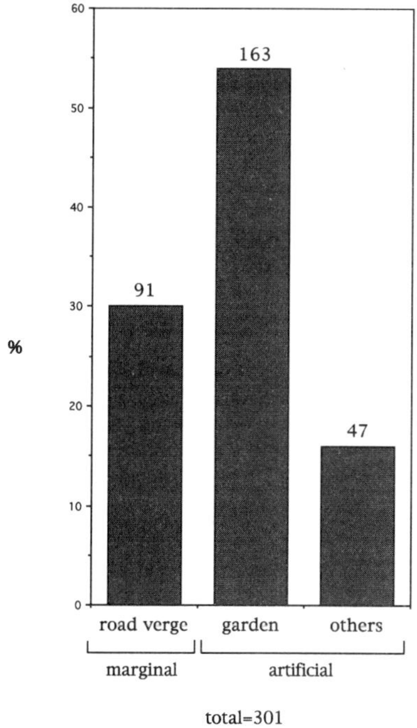

Fig. 1.5. Hedgehog: habitats

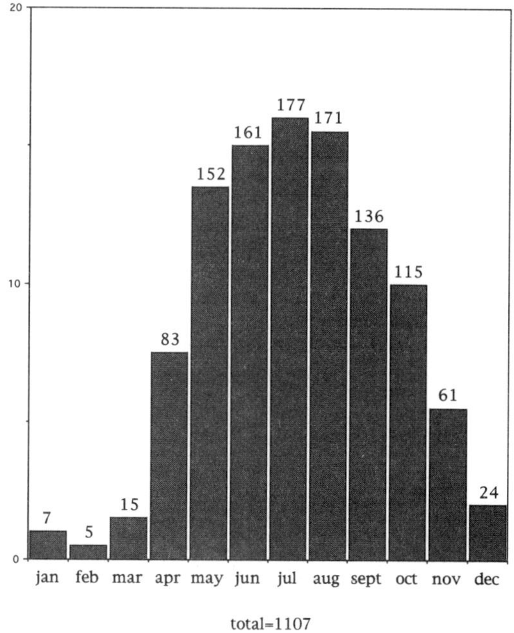

Fig. 1.6. Hedgehog: mortality (road casualties by month)

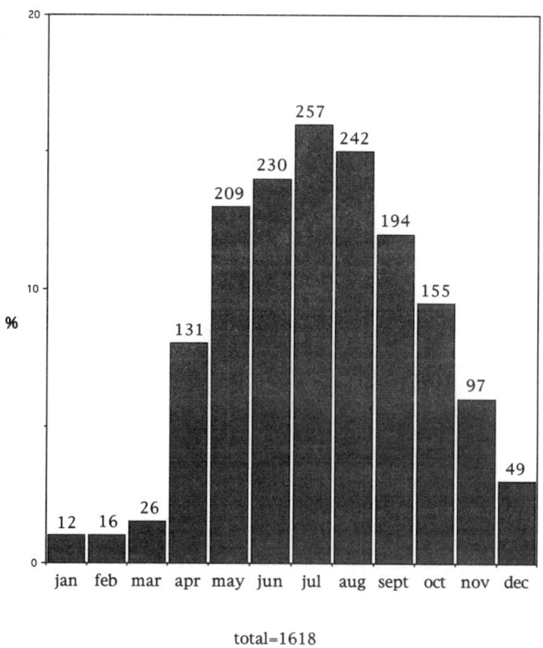

Fig. 1.7. Hedgehog: seasonal activity

Order Insectivora — Family Talpidae

Mole — *Talpa europaea*

Easily recognisable insectivore adapted for predominantly subterranean existence with velvety, black fur and large front feet for digging. Head and body 110-160 mm. Foot pads (fig. 2.1), skull, jawbones and dentition (fig. 2.2) are diagnostic.

Handbook, 43-49; *Atlas*, 26-27; *Review*, 9-10.

Status

Native. Common.

The total pre-breeding population in Britain is estimated at 31,000,000. Population trends are unknown.

Nationally

Widespread distribution, found in a wide range of habitats and where the soil is sufficiently deep for them to dig tunnels.

Wiltshire

Most historical records are of the mole as an agricultural pest. Mole-catching was paid for out of the church rates at Yatesbury from 1792 at a fixed sum of £2.12s.6d per annum (Smith 1879). In 1908, an increase in moles on the Marlborough Downs to the extent that "mole hills predominate over the turf", was put down to lack of trapping (*MCNHS Report* 57). Of earlier writers, Smith (1875) alone made a case for not persecuting the mole. His article in *WAM* covered many aspects of the natural history and folklore of the species. The extent to which the species is still persecuted is not known. A recent study revealed that in neighbouring Dorset, 49.2% of 120 farmers surveyed reported damage and 45.1% (of 122 respondents) claimed to control moles (Atkinson, Macdonald and Johnson 1994).

Special appeal for records made 1986. Field surveys conducted 1986-87 in 10 km squares which were under-recorded. Data up to 1987 published in *WAM* 82, 147-155.

1008 records, of which 985 have been used in this analysis. 90% of records are of field signs (fig. 2.3). Widespread distribution, particularly in river valleys (fig. 2.4), in a wide range of habitats, especially farmland (fig. 2.5); probably under-recorded on the chalklands. Records of mortality are mostly of casual finds; there are comparatively few records of trapping and predation (fig. 2.6).

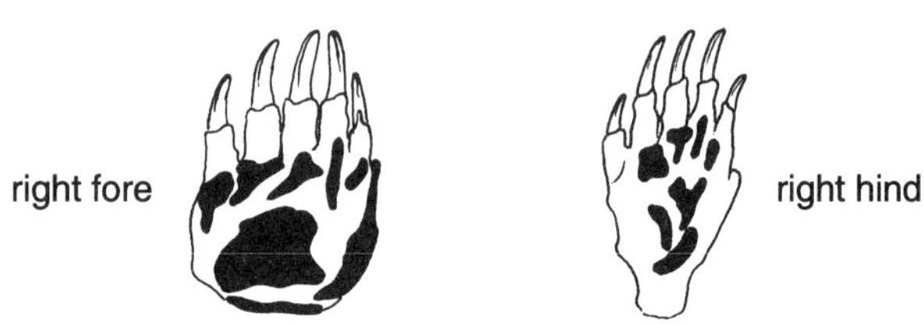

Fig. 2.1. Mole: foot pads, approximately x 1

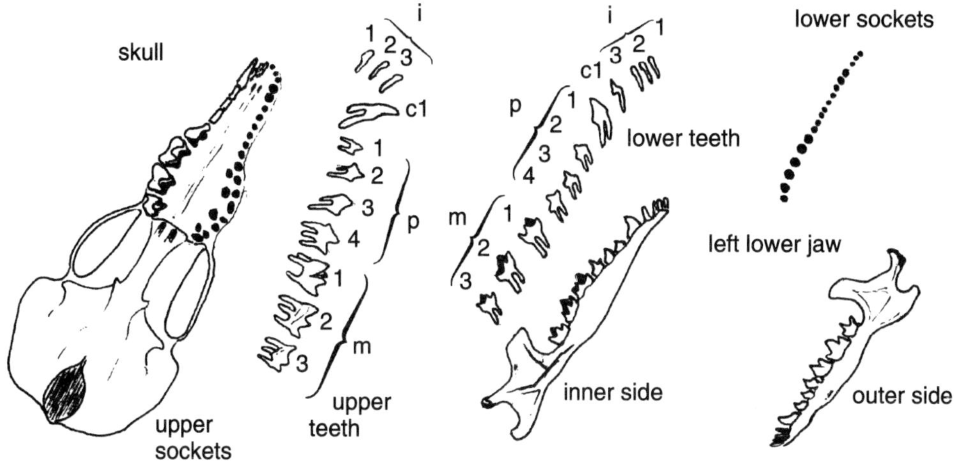

Fig. 2.2. Mole: skull, jawbones and dentition, approximately x 1.
i, incisor; c, canine; p, premolar; m, molar

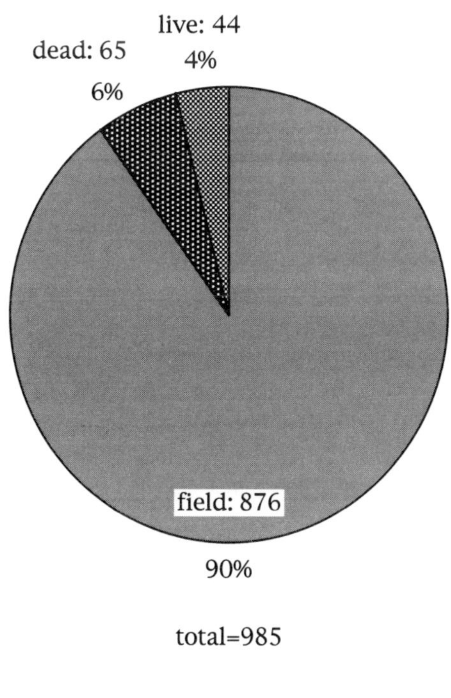

Fig. 2.3. Mole: number and nature of record

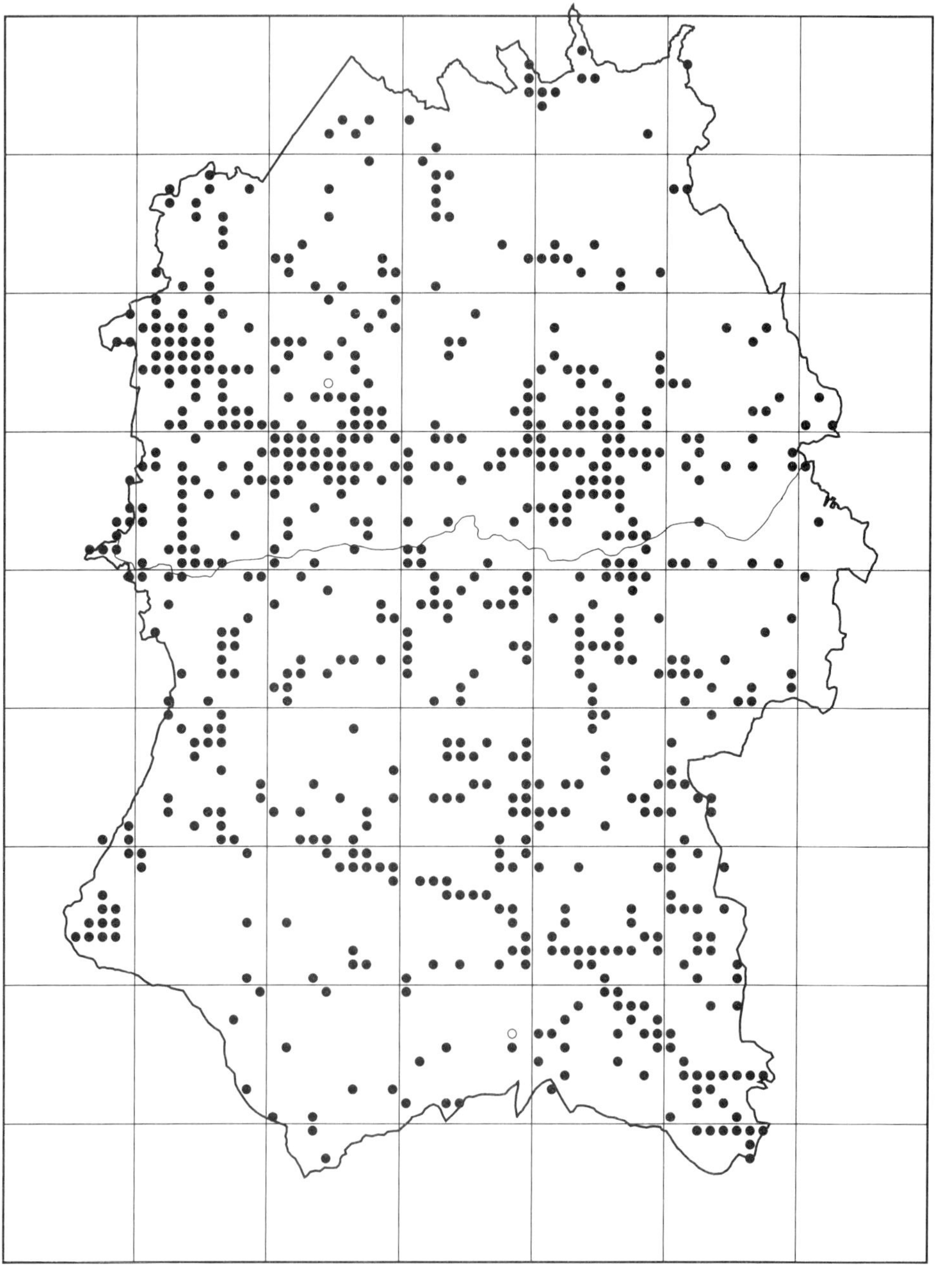

Fig. 2.4. Mole: distribution

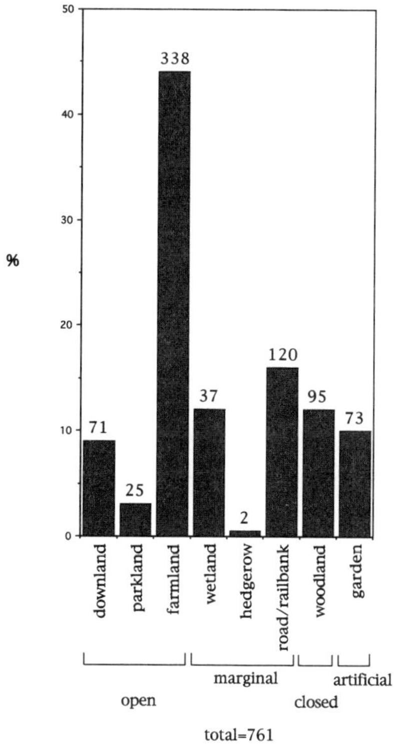

Fig. 2.5. Mole: habitats

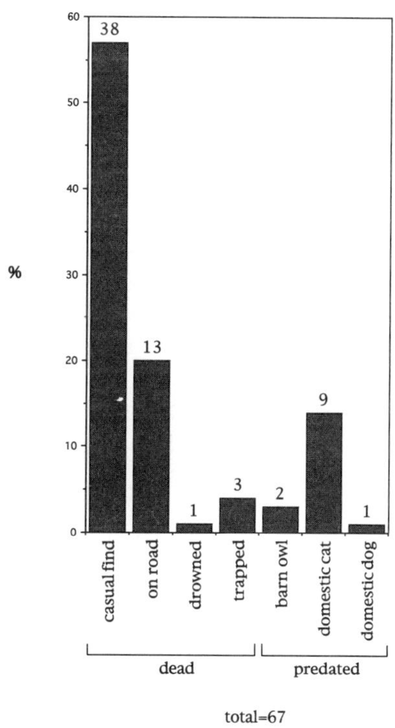

Fig. 2.6. Mole: mortality

Order Insectivora — Family Soricidae

Common shrew — *Sorex araneus*

Small animal with very long pointed muzzle and a tail approximately the same length as the body. The fur is velvety and very dense, dark grey or brownish grey on the upper parts of the body, paler on the underparts. Head and body 48-80 mm. Larger than pygmy shrew but smaller than water shrew (fig. 3.1). Skulls, jawbones and dentition for all three species are diagnostic (fig. 3.2).

Handbook, 51-58; *Atlas*, 26-27; *Review*, 10-11; *Red Data Book*, 14-16.

Status

Native. Common.

The total pre-breeding population in Britain is estimated at 41,700,000. Population trends are unknown; there may be a decline through habitat loss and the use of insecticides. Agricultural 'set-aside' may lead to some increase locally.

Nationally

Widespread distribution; found in a wide range of habitats, particularly those which offer good ground cover.

Wiltshire

Special appeal for records made 1983-85. Data up to 1985 published in *WAM* 80, 197-210.

365 records of which 344 have been used in this analysis. 63% of the records are of dead animals (fig. 3.3). Widespread but patchy distribution (fig. 3.4) in a wide range of habitats, particularly gardens and road- and railbanks (fig. 3.5). Patchiness of distribution is probably due to under-recording. Field signs are mainly of skeletal material in bird pellets. Wide range of mortality factors, particularly casual finds of dead animals and predation by domestic cat (fig. 3.6).

Fig. 3.1: Comparative sizes and proportions of shrews, approximately x ½

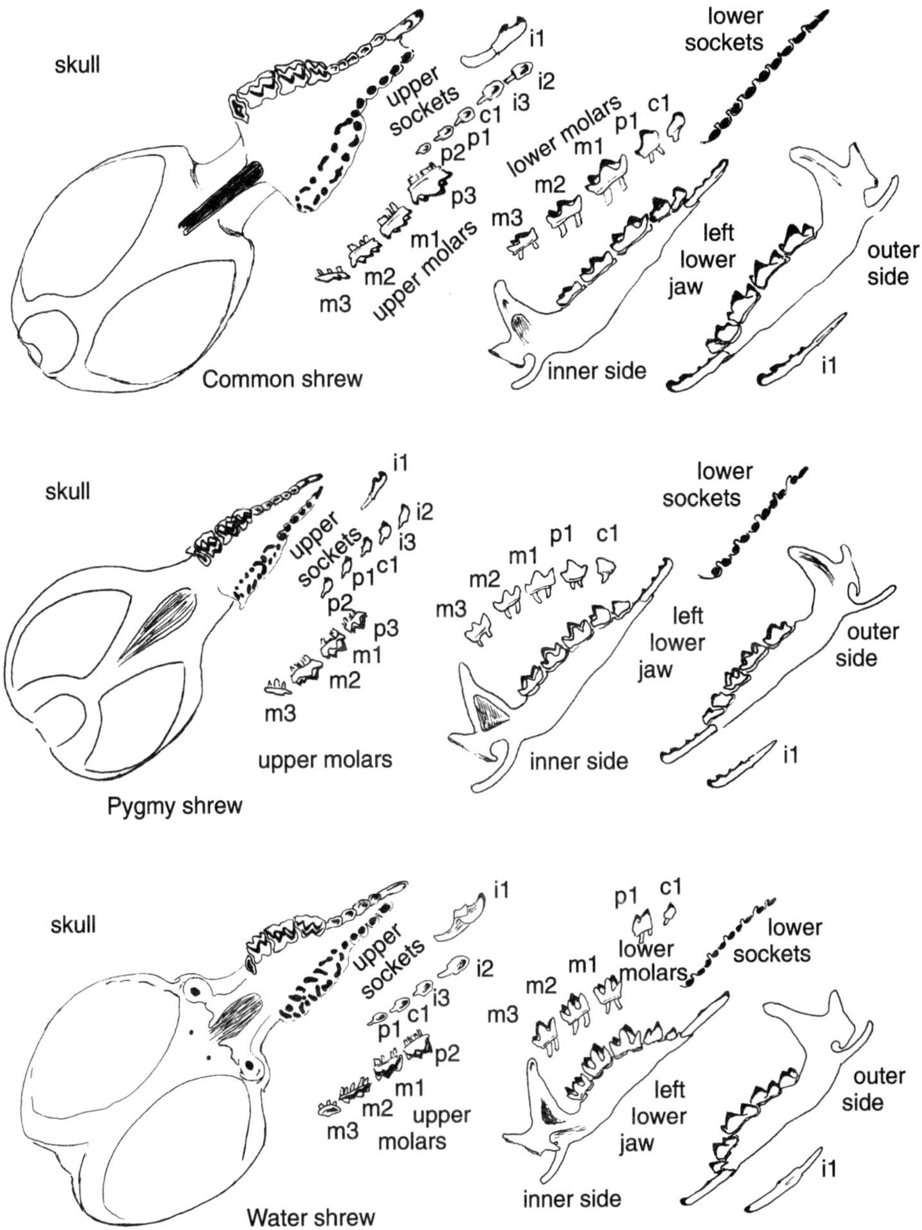

Fig. 3.2: Skulls, jawbones and dentition of shrews, approximately x 2.
Top: common shrew; centre: pygmy shrew; bottom: water shrew

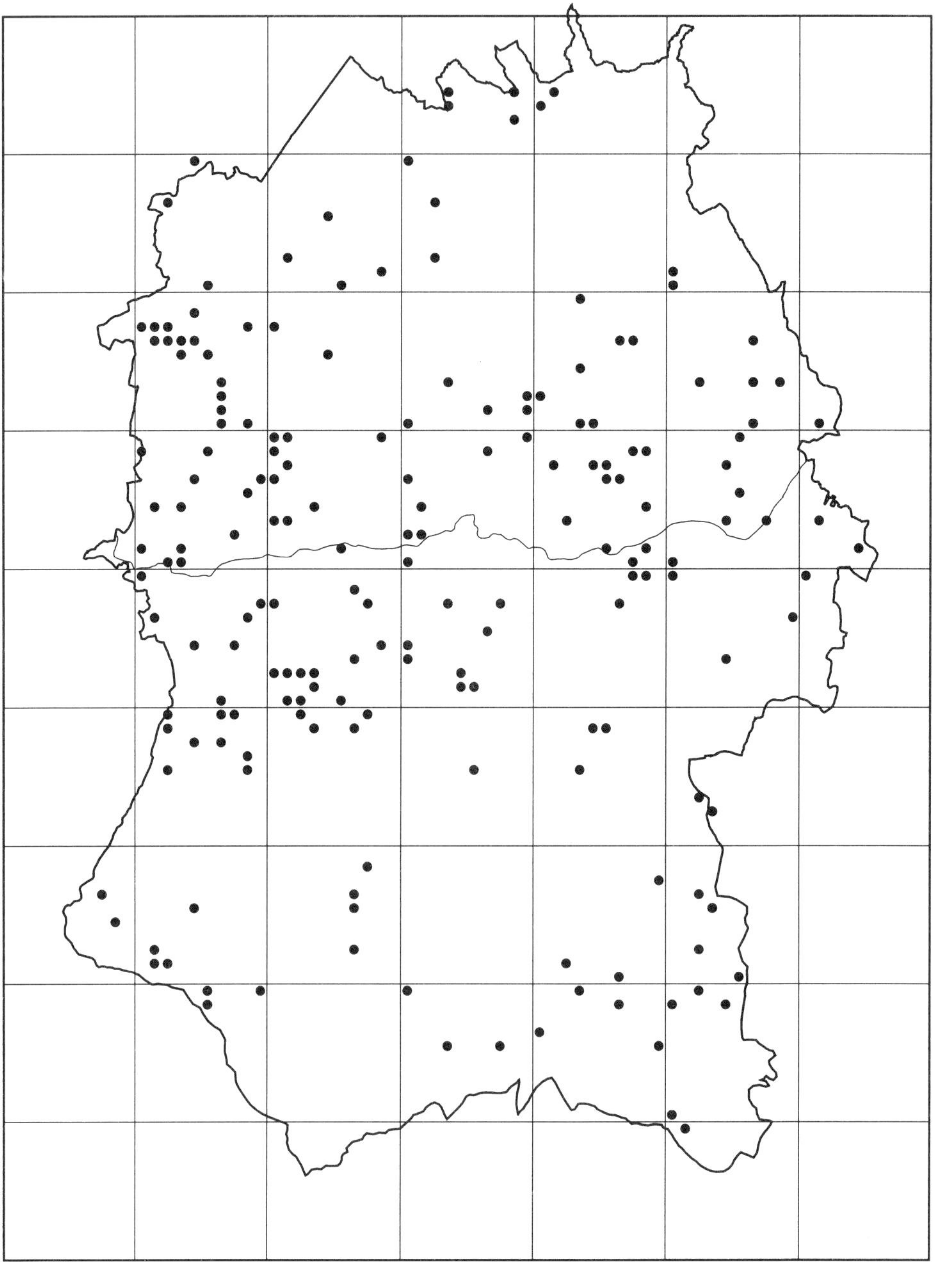

Fig. 3.4. Common shrew: distribution

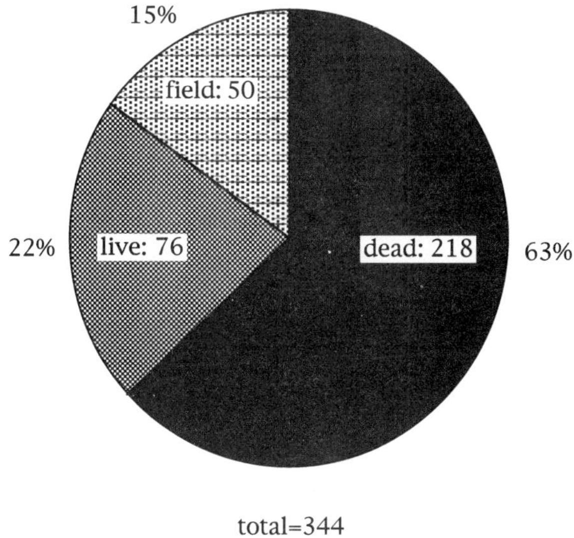

Fig. 3.3. Common shrew: number and nature of record

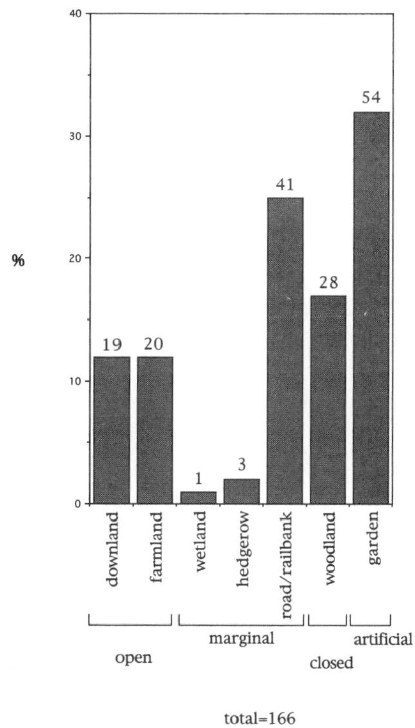

Fig. 3.5. Common shrew: habitats

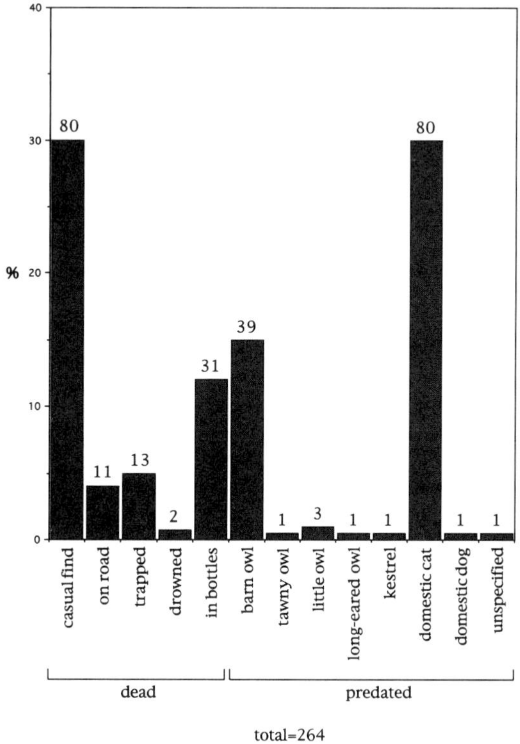

Fig. 3.6. Common shrew: mortality

Order Insectivora — Family Soricidae

Pygmy shrew — *Sorex minutus*

Small animal with very long pointed muzzle and a tail approximately the same length as the body. The fur is velvety and very dense, dark grey or brownish grey on the upper parts of the body, paler on the underparts. Head and body 40-60 mm. The smallest British shrew (fig. 3.1). Skull, jawbones and dentition are diagnostic (fig. 3.2).

Handbook, 60-64; *Atlas*, 28-29; *Review*, 11-12; *Red Data Book*, 14-16.

Status

Native. Common.

The total pre-breeding population in Britain is estimated at 8,600,000. Population trends are unknown; there may be a decline through habitat loss and the use of insecticides. Agricultural 'set-aside' may lead to some increase locally.

Nationally

Widespread distribution; found in a wide range of habitats, particularly those which offer good ground cover.

Wiltshire

Special appeal for records made 1983-85. Data up to 1985 published in *WAM* 80, 197-210.

177 records of which 169 have been used in this analysis. 54% of the records are of dead animals (fig. 4.1). Widespread but very patchy distribution (fig. 4.2). The patchiness of distribution is probably due to under-recording. A more restricted range of habitats recorded than for the common shrew (fig. 4.3). All 42 records of field signs are of skeletal material in bird pellets. Smaller range of mortality factors than for common shrew but nevertheless still comparatively wide (fig. 4.4); the apparent predominance of predation by domestic cat and barn owl is probably due to outcomes of specific surveys ('what the cat brought in' (Ward 1981) and owl pellet workshops (see, for example, Dillon 1977)).

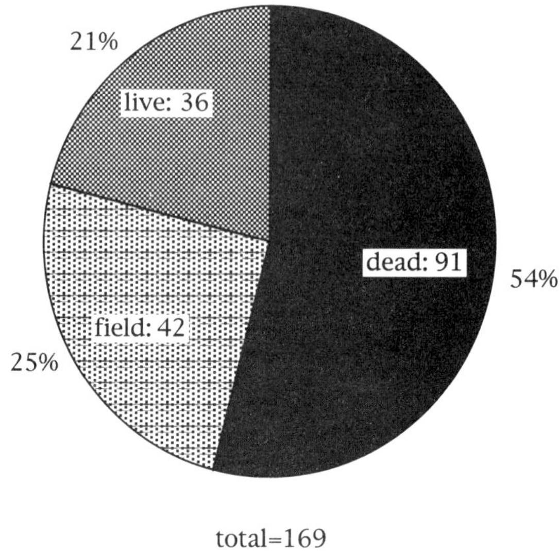

Fig. 4.1. Pygmy shrew: number and nature of record

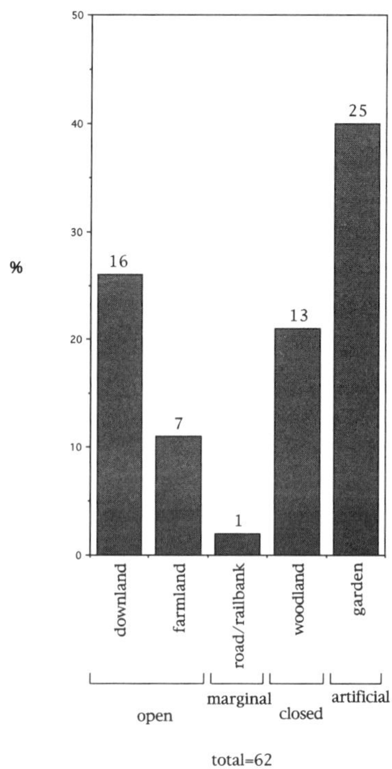

Fig. 4.3. Pygmy shrew: habitats

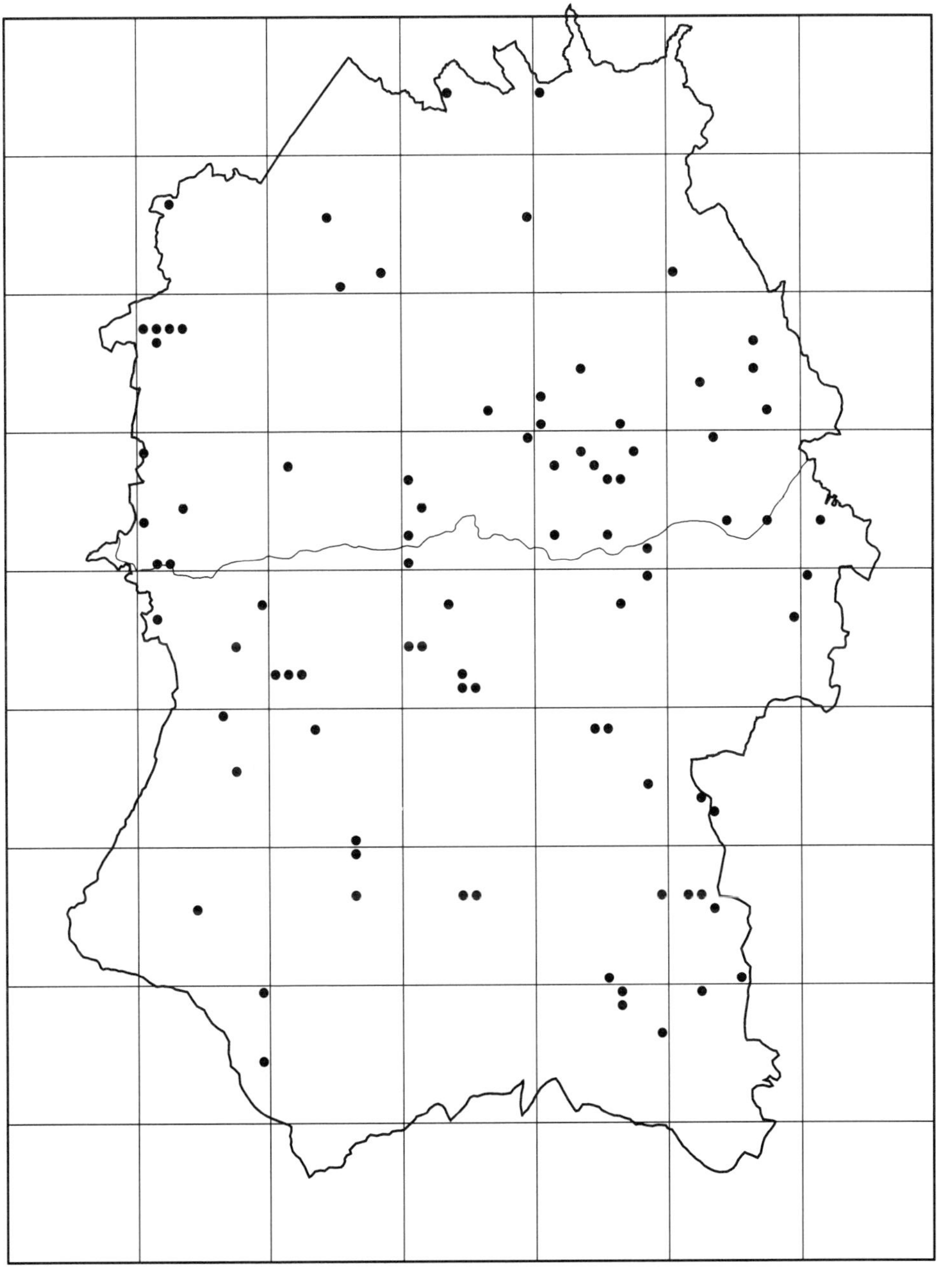

Fig. 4.2. Pygmy shrew: distribution

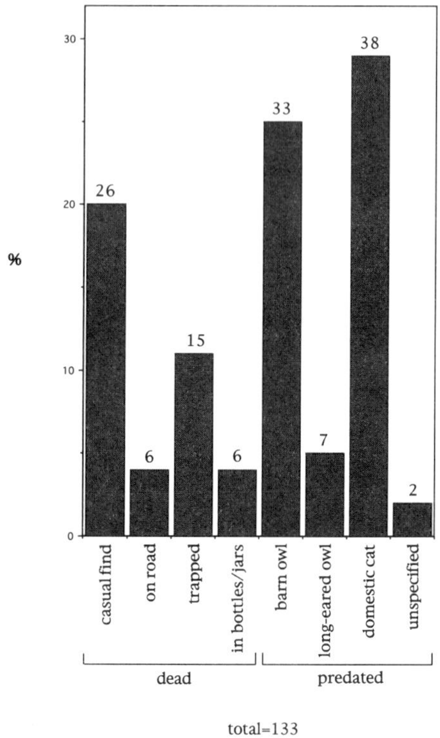

Fig. 4.4. Pygmy shrew: mortality

Order Insectivora — Family Soricidae

Water shrew — *Neomys fodiens*

Small animal with very long pointed muzzle and a tail approximately the same length as the body. The fur is velvety and very dense, with a contrast between the dark, sometimes black, upper and the pale, sometimes white underparts. Head and body 67-96 mm. The largest of the three shrews found in Wiltshire (fig. 3.1). Skull, jawbones and dentition are diagnostic (fig. 3.2).

Handbook, 64-68; *Atlas*, 30-31; *Review*, 12-13; *Red Data Book*, 14-16.

Status

Native. Common where it occurs.

The total pre-breeding population in Britain is estimated at 1,900,000. Population trends are unknown.

Nationally

Widespread distribution; generally associated with wetland habitats but may be found a considerable distance from water.

Wiltshire

Special appeal for records made 1983-85. Data up to 1985 published in *WAM* 80, 197-210.

112 records of which 98 have been used in this analysis. In contrast to both the common and pygmy shrews, the majority (60%) of the records are of live animals (fig. 5.1). Widespread but very patchy distribution with concentrations of records on some of the river systems in the county (fig. 5.2). A more restricted range of habitats than for the common shrew and dominated by wetland (fig. 5.3). Apparent absence, where habitat is suitable, is probably due largely to under-recording. All six records of field signs are of skeletal material in bird pellets. A smaller range of mortality factors than for both common and pygmy shrews, dominated by casual finds of dead animals (fig. 5.4).

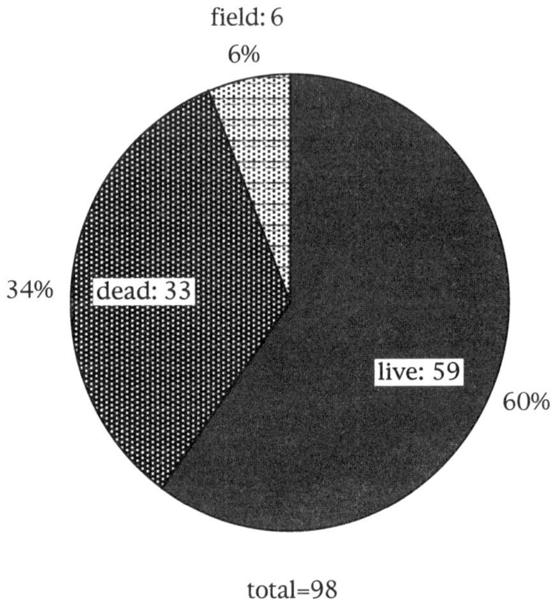

Fig. 5.1. Water shrew: number and nature of record

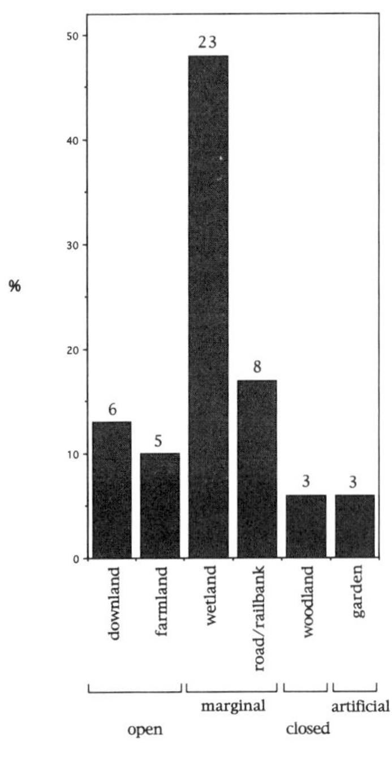

Fig. 5.3. Water shrew: habitats

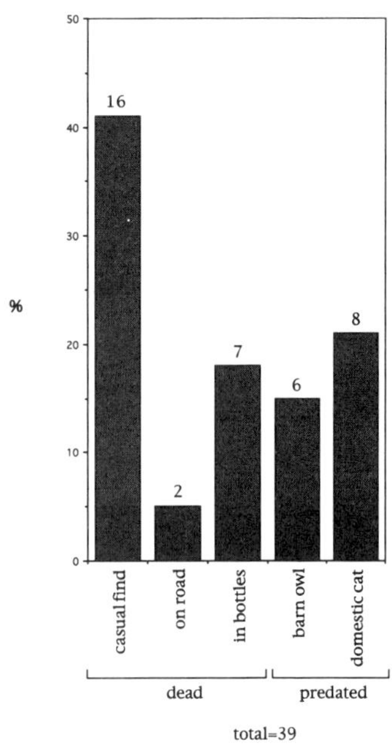

Fig. 5.4. Water shrew: mortality

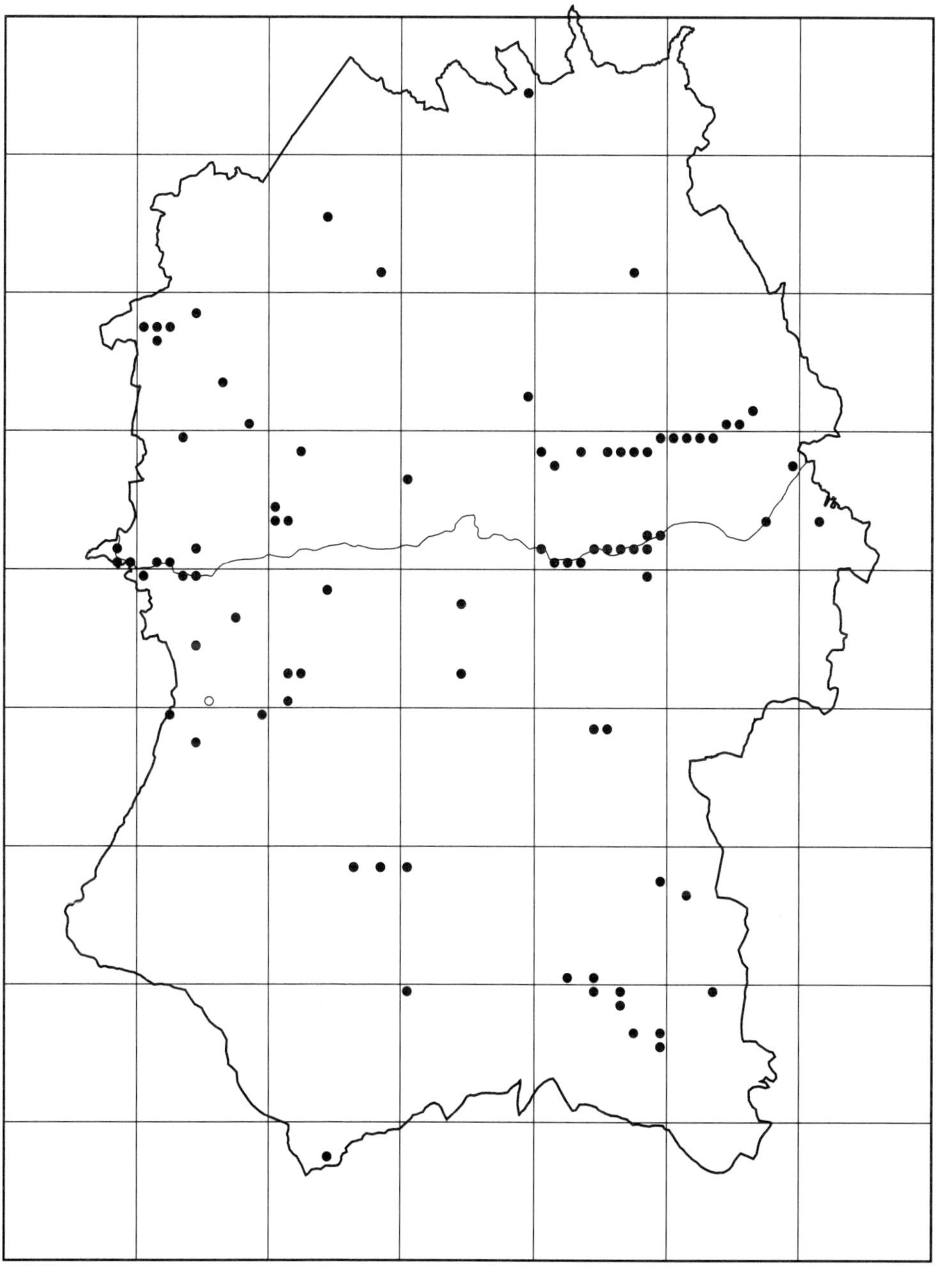

Fig. 5.2. Water shrew: distribution

Order Chiroptera — Family Rhinolophidae

Greater horseshoe bat — *Rhinolophus ferrumequinum*

A 'tablespoon' sized bat and one of two British species with a distinctive horseshoe-shaped 'nose leaf'. The other is the much smaller lesser horseshoe bat. Head and body 58-68 mm; wingspan 330-395 mm.

Handbook, 88-94; *Atlas,* 34-36; *Review,* 17-19; *Red Data Book,* 17-23.

Status

Native. Endangered.

The total pre-breeding population in Britain is estimated at 4,000. Status in historic times unknown; there has been a decline in recent decades. Numbers in some colonies may be stabilising as a result of conservation measures.

Nationally

Limited and localised distribution in southern England and Wales, mainly in the south-west and dependent on the availability of suitable roosting sites – caves, mines, quarries and (in summer) buildings.

Wiltshire

Earliest published record 1910 (Hony 1917).

A general appeal for records was made in 1982 and, in 1985, a survey of churches was launched, but no detailed account of the bats has been published.

41 records of which 37 have been used in this analysis. 95% of the records are of live animals. Very limited and localised distribution in the south and west of the county (fig. 6.1), mainly associated with roosting sites, of which domestic buildings (12), quarries (9), and churches (4) have been recorded. The two records of field signs are of droppings. No information about mortality factors.

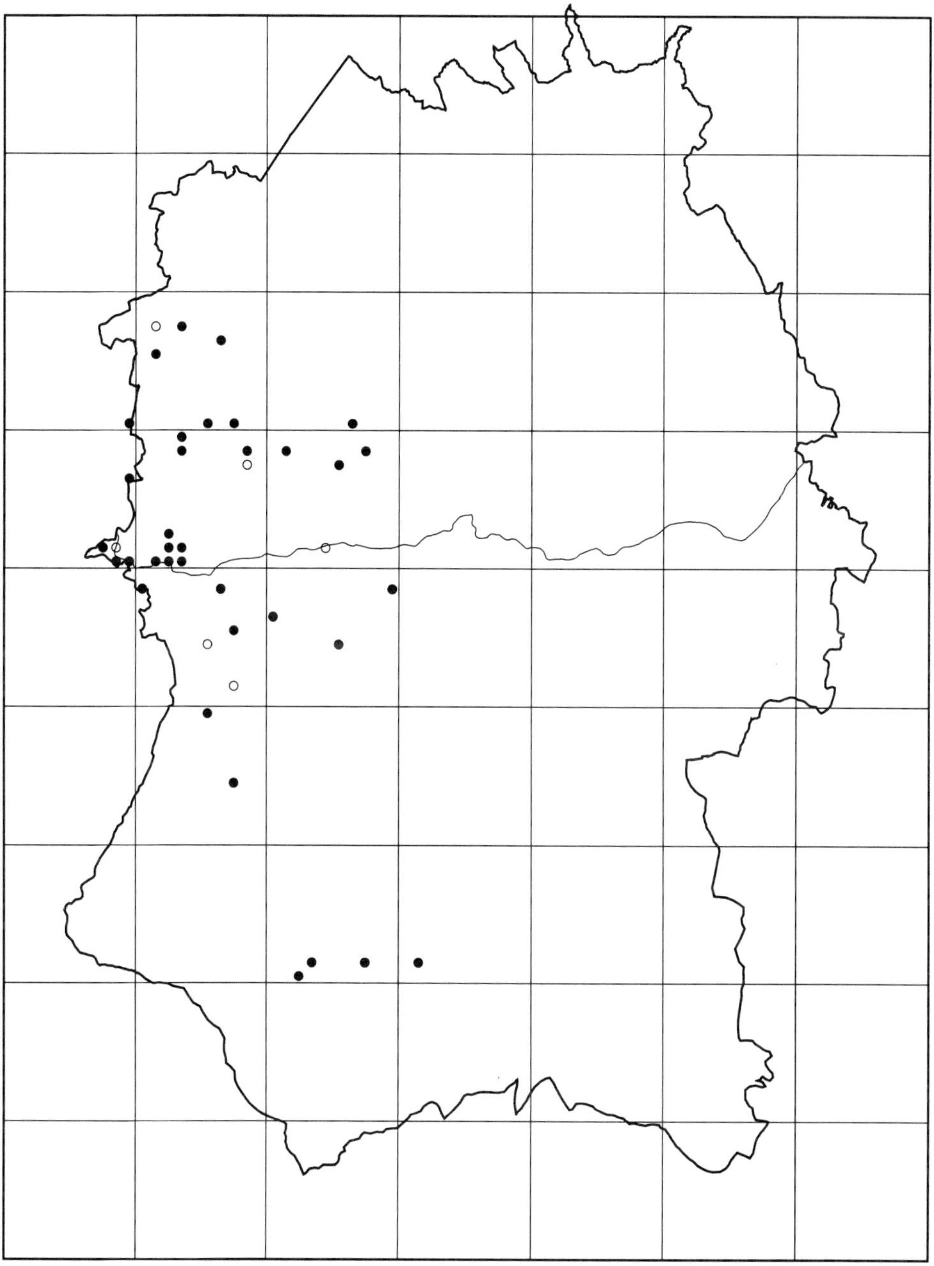

Fig. 6.1. Greater horseshoe bat: distribution

Order Chiroptera — Family Rhinolophidae

Lesser horseshoe bat — *Rhinolophus hipposideros*

A 'teaspoon' sized bat and one of two British species with a distinctive horseshoe-shaped 'nose leaf'. The other is the much larger greater horseshoe bat. Head and body 35-39 mm; wingspan 225-250 mm.

Handbook, 95-98; *Atlas,* 36-37; *Review,* 19-20; *Red Data Book,* 17-23.

Status

Native. Endangered.

The total pre-breeding population in Britain is estimated at 14,000. Status in historic times unknown; there has been a decline in recent decades. Numbers in some colonies may be stabilising as a result of conservation measures.

Nationally

More widespread distribution in south-west England and Wales than the greater horseshoe bat but dependent on the availability of suitable roosting sites – caves, mines, quarries and buildings – often associated with woodland or scrub.

Wiltshire

A specimen taken by Montagu in a hollow over a baker's oven at Lackham was reported in the *Transactions of the Linnean Society of London* in 1804. According to Bell (1837) and Hony (1917) this is the first English record of the species.

A general appeal for records was made in 1982 and, in 1985, a survey of churches was launched, but no detailed account of the bats has been published.

34 records, all of which have been used in this analysis. 94% of the records are of live animals. Very limited and localised distribution in the south and west of the county (fig. 7.1), mainly associated with roosting sites, of which quarries (12), domestic buildings (8), and 'other' buildings (3) have been recorded. The two records of field signs are of droppings. No information about mortality factors.

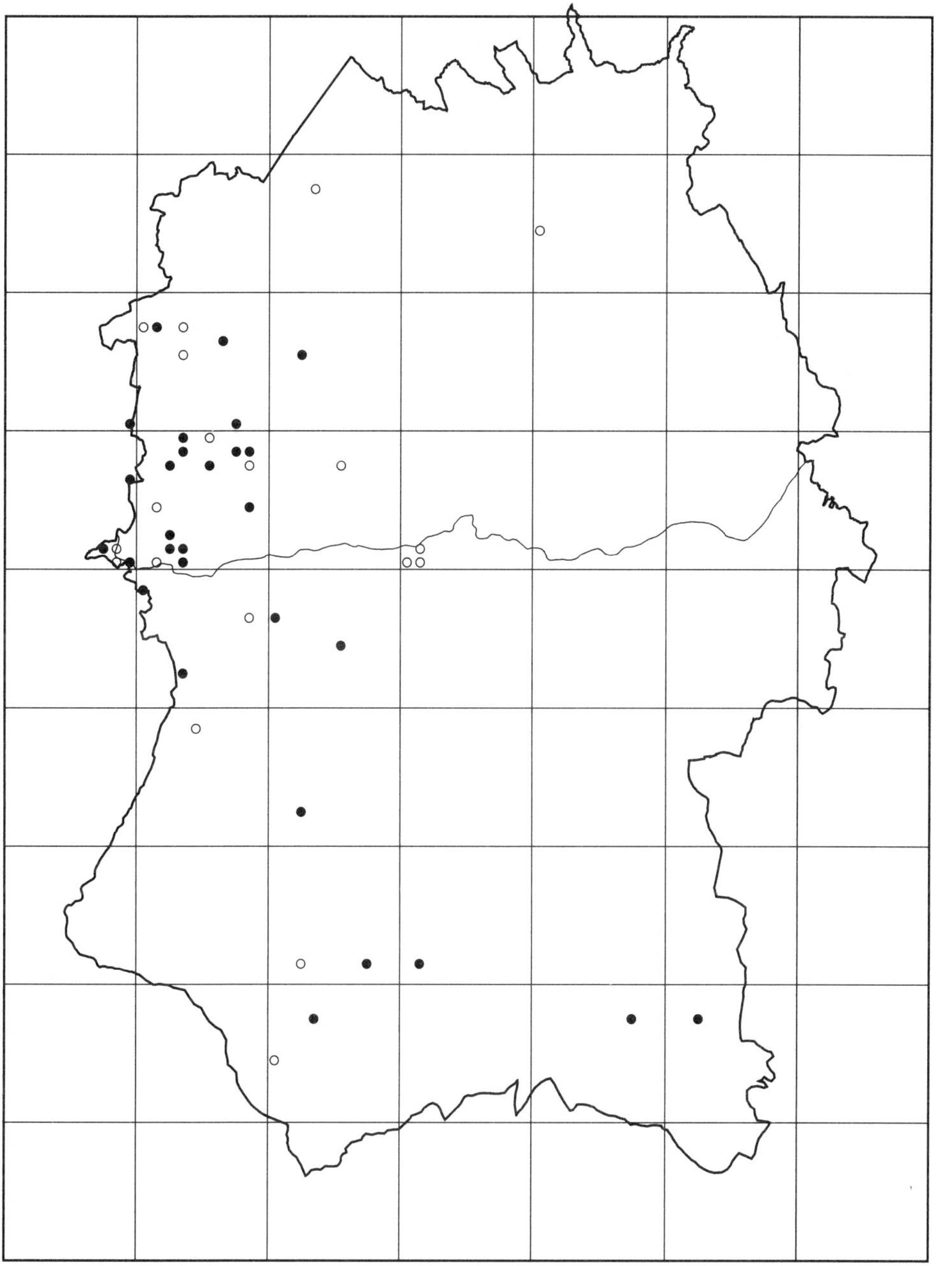

Fig. 7.1. Lesser horseshoe bat: distribution

Order Chiroptera — Family Vespertilionidae

Whiskered bat — *Myotis mystacinus*

A small bat with dark skin and small ears which can be distinguished from Brandt's bat by differences in the ear, tragus (a lobe developed from the lower rim of the ear) and teeth. Head and body 35-48 mm; wingspan 210-240 mm.

Handbook, 99-101; *Atlas*, 40-41; *Review*, 21; *Red Data Book*, 17-23.

Status

Native. Vulnerable.

The total pre-breeding population in Britain is estimated at 40,000. Status in historic times unknown; there has been a decline in recent decades.

Nationally

Widely but sparsely distributed in England and Wales in both wooded and open habitats, often near water.

Wiltshire

The earliest record of this species is of a specimen taken at Alton Barnes in 1897 and now in the British Museum (Natural History) (Hony 1917).

A general appeal for records was made in 1982 and, in 1985, a survey of churches was launched, but no detailed account of the bats has been published.

40 records of which 39 have been used in this analysis. 77% of the records are of live and 23% are of dead animals. Very limited and localised distribution in the west and south of the county (fig. 8.1), mainly associated with roosting sites in quarries (12), and domestic and 'other' buildings (8); there is only one record from a church. Of the nine records of dead animals, eight are of casual finds and one is of a drowned specimen.

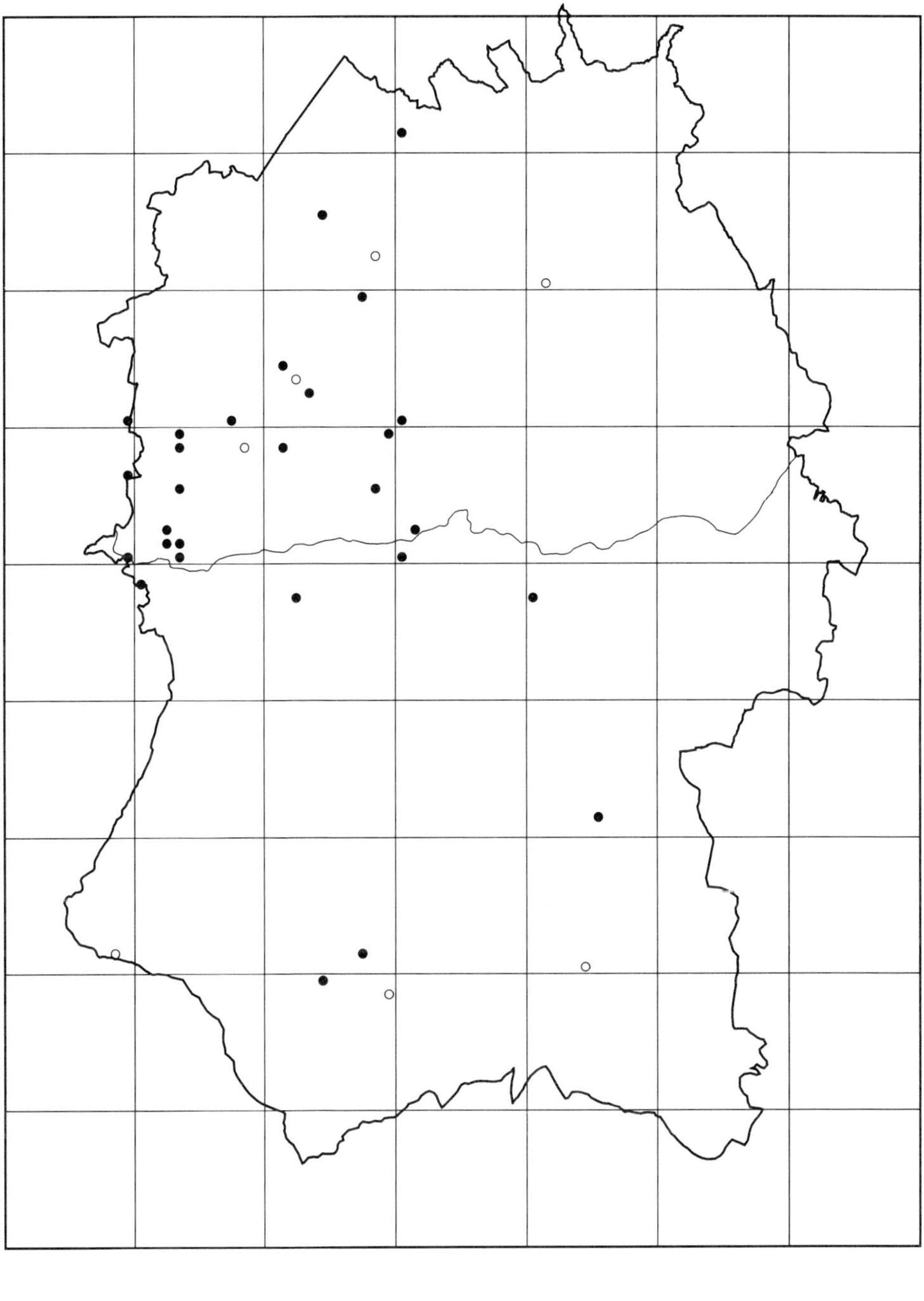

Fig. 8.1. Whiskered bat: distribution

Order Chiroptera — Family Vespertilionidae

Brandt's bat — *Myotis brandtii*

A small bat with slightly redder upper fur than the whiskered bat but otherwise distinguished from it by differences in the ear, tragus (a lobe developed from the lower rim of the ear) and teeth. Head and body 37-48 mm; wingspan 210-255 mm.

Handbook, 101-102; *Atlas,* 42-43; *Review,* 21-22; *Red Data Book,* 17-23.

Status

Native. Vulnerable.

The total pre-breeding population in Britain is estimated at 30,000. Probably declining.

Nationally

Widely distributed in England and Wales but more sparsely than the whiskered bat. Usually associated with wooded habitats and roosting sites (mainly caves).

Wiltshire

A general appeal for records was made in 1982, and in 1985, a survey of churches was launched, but no detailed account of the bats has been published.

Seven records, all of which have been used in this analysis. All the records are of live animals. Very limited and localised distribution in the west and south of the county (fig. 9.1), mainly associated with roosting sites in buildings and quarries. No information on mortality.

Until 1970, whiskered and Brandt's bats were regarded as one species. In addition to the records held for the separate species, the Wiltshire Biological Records Centre holds 21 undifferentiated records. The undifferentiated records have not been used in this analysis.

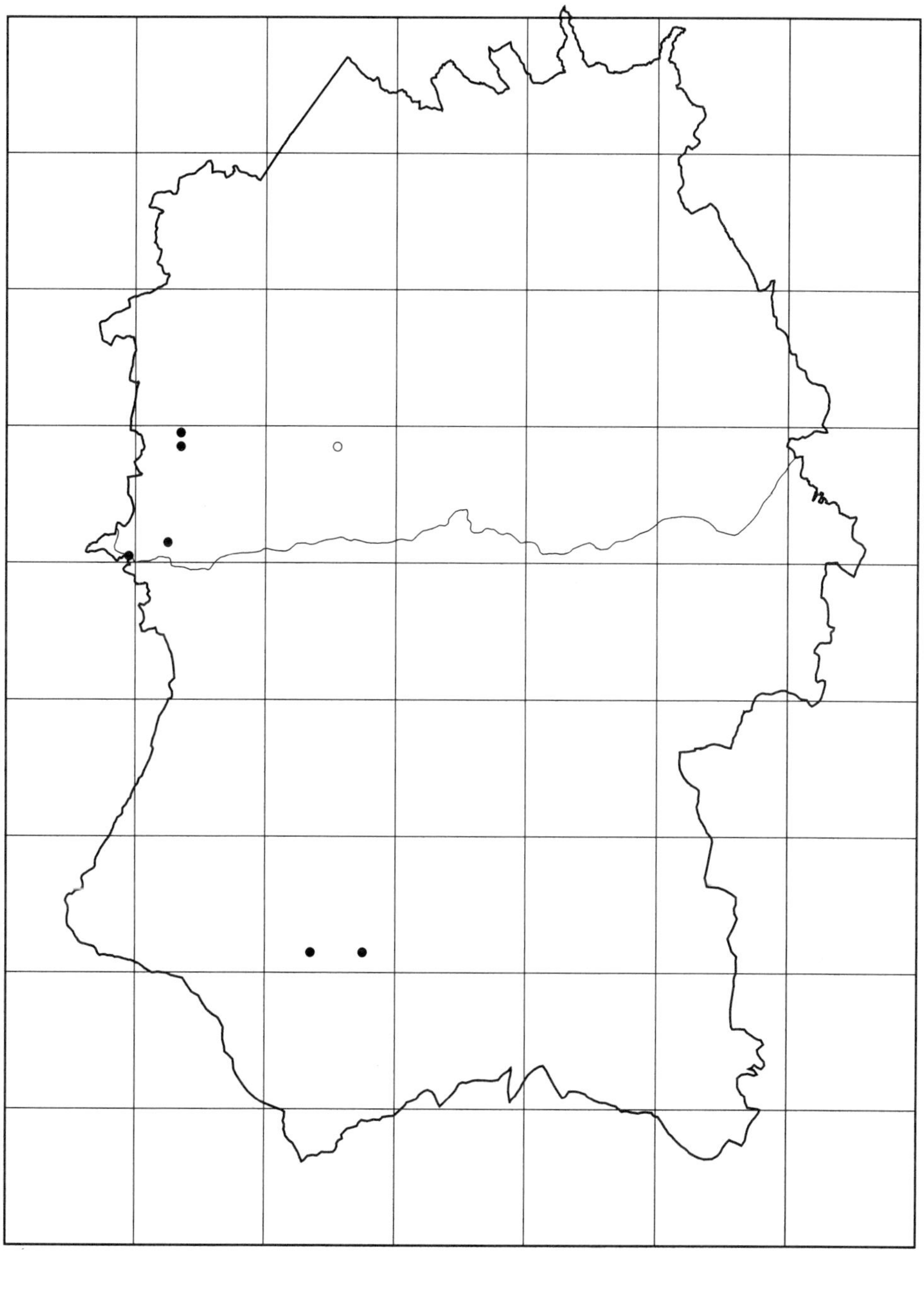

Fig. 9.1. Brandt's bat: distribution

Order Chiroptera — Family Vespertilionidae

Natterer's bat — *Myotis nattereri*

A medium sized bat with a paler face than the whiskered and Brandt's bats. The distinctive feature of Natterer's bat is a fringe of stiff hairs along the outer edge of the tail membrane from the tail to the end of the calcar (a bony rod arising from the ankle and supporting the trailing edge of the tail membrane). Head and body 40-50 mm; wingspan 250-300 mm.

Handbook, 102-105; *Atlas*, 44-45; *Review*, 22-23; *Red Data Book*, 17-23.

Status

Native. Views about status vary from vulnerable (*Atlas*) to fairly common (*Review*).

The total pre-breeding population in Britain is estimated at 100,000. Status in historic times and present population trends unknown.

Nationally

Widely distributed in England, Wales and Scotland and often associated with roosts in wooded and parkland habitats.

Wiltshire

There used to be a specimen in the Blackmore Museum, Salisbury, dated 1859. Although without provenance, it was thought to have been taken locally (Hony undated). The first provenanced record is of a number of specimens in the Chilmark Quarries in 1916 (Hony undated).

A general appeal for records was made in 1982 and, in 1985, a survey of churches for signs of bats was launched, but no detailed account of the bats has been published.

67 records of which 62 have been used in this analysis. 66% of the records are of live and 31% are of dead animals; the remaining 3% are of field signs. Scattered distribution in Wiltshire, recorded mainly from the west and south of the county (fig. 10.1), mainly associated with roosting sites in domestic and 'other' buildings (18) and quarries (10). Probably under-recorded. The two records of field signs are of droppings. Of the 19 records of dead animals, 15 are of casual finds and four are of animals killed by domestic cats.

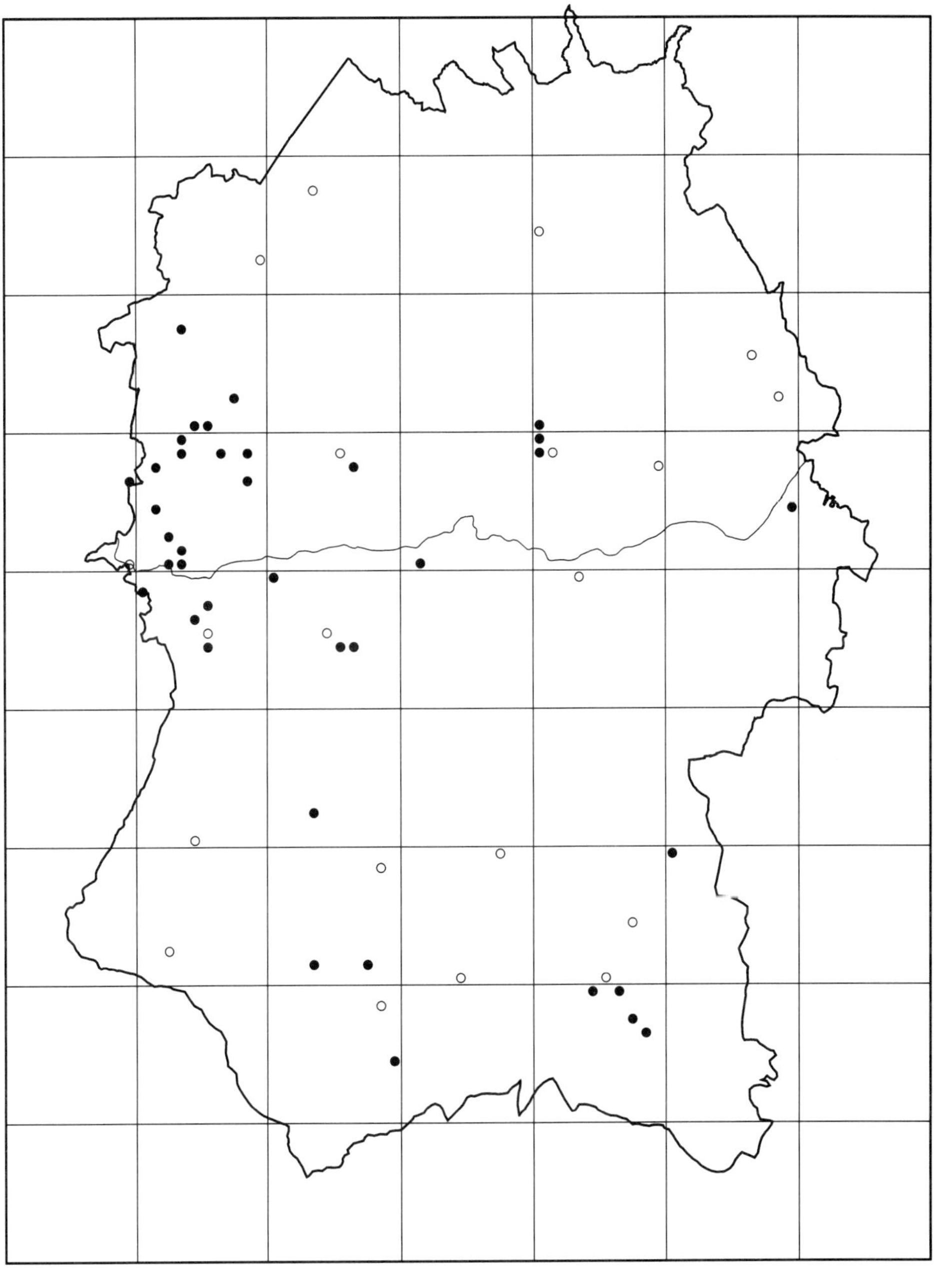

Fig. 10.1. Natterer's bat: distribution

Order Chiroptera — Family Vespertilionidae

Bechstein's bat — *Myotis bechsteinii*

A medium sized bat, similar to Natterer's bat but with large ears which, unlike the long-eared bats, do not meet at the base. Head and body 43-50 mm; wingspan 250-300 mm.

Handbook, 105-107; *Atlas,* 46-47; *Review,* 23; *Red Data Book,* 17-23.

Status

Native. Rare.

The total pre-breeding population in Britain is estimated at 1,500. Status in historic times unknown. Present population trends unknown although the species may be vulnerable to habitat fragmentation.

Nationally

Limited and localised distribution in woodland in southern and western England.

Wiltshire

First record attributed to Dr. Blackmore who told R.S. Newall that he caught one in Salisbury in the early years of the century (*MS* letter to G.B. Hony, c.1916, WANHS).

A general appeal for records was made in 1982 and, in 1985, a survey of churches was launched, but no detailed account of the bats has been published.

Nine records, all of which have been used in this analysis. Six of the records are of live animals, one is of a dead animal, and two are of field signs. Very localised distribution in the south-west of the county (fig.11.1). A third of the records are associated with roosting sites in buildings. Possibly under-recorded. The two records of field signs are of droppings. The dead animal was killed by a domestic cat.

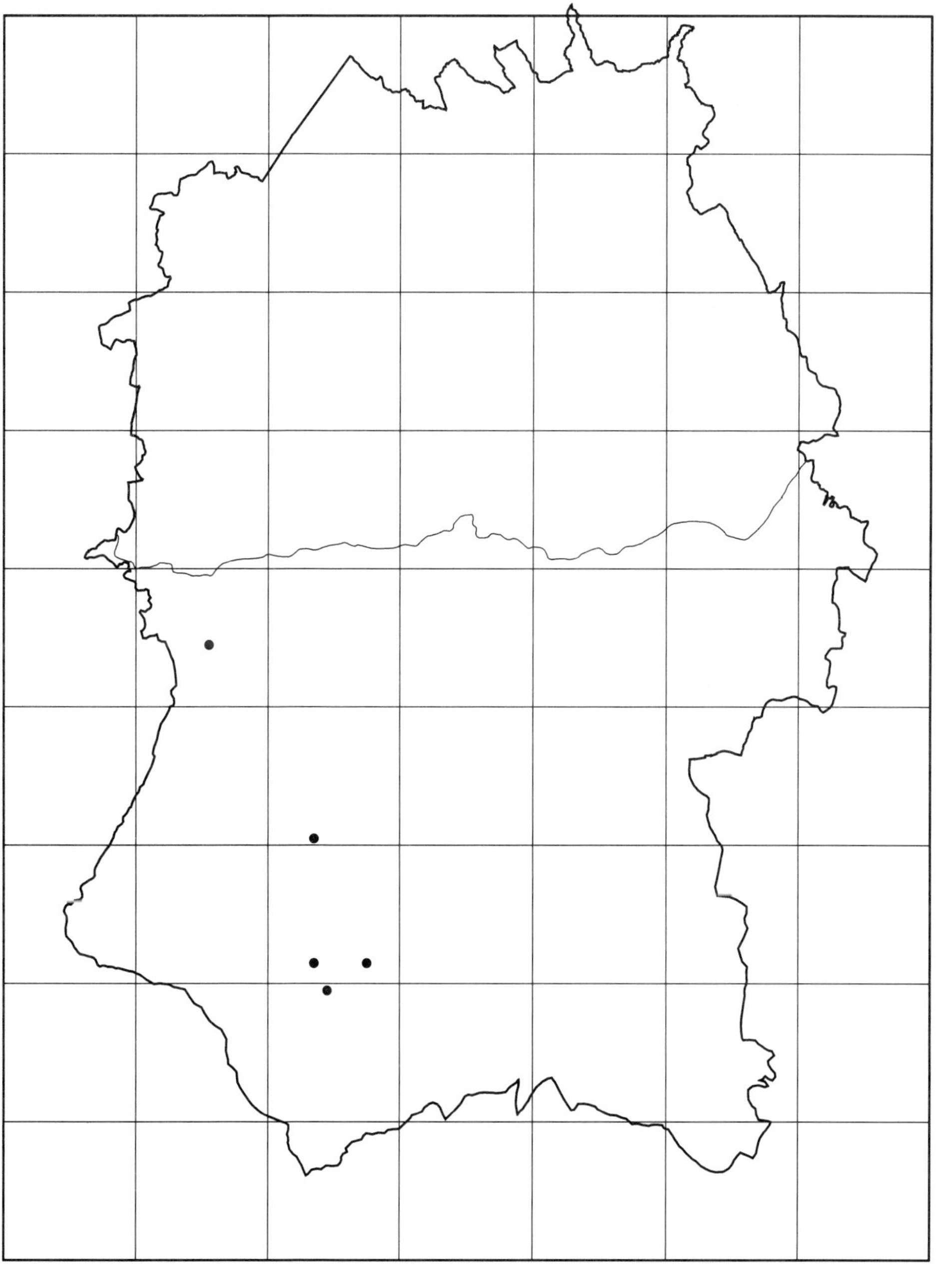

Fig. 11.1. Bechstein's bat: distribution

Order Chiroptera — Family Vespertilionidae

Daubenton's bat — *Myotis daubentonii*

A small bat with even-textured, mole-like fur, a long calcar (a bony protrudence from the ankle which supports the trailing edge of the tail membrane) and large feet. Head and body 40-50 mm; wingspan 230-270 mm.

Handbook, 108-111; *Atlas,* 50-51; *Review,* 24-25; *Red Data Book,* 17-23.

Status

Native. Views about status vary from vulnerable (*Atlas*) to common (*Review*).

The total pre-breeding population in Britain is estimated at 150,000. Status in historic times and present population trends unknown.

Nationally

Widely distributed in England, Wales and Scotland. Often associated with wetlands in combination with woodland and parkland habitats.

Wiltshire

First record is of a specimen in a hollow tree at Lydiard Park, 1916 (Hony undated). This record pre-dates the claim made in 1957 for the specimen found in Fonthill grottos (*WAM* 57, 146).

A general appeal for records was made in 1982 and, in 1985, a survey of churches was launched, but no detailed account of the bats has been published.

52 records of which 41 have been used in this analysis. 66% of the records are of live and 24% are of dead animals; the remaining 10% are of field signs. Scattered distribution in Wiltshire (fig. 12.1). Of the 27 habitat records, 12 are of wetland, six are of quarries, four are of churches, three are of buildings and two are of gardens. Probably under-recorded. One of three bat species to be successfully identified in Wiltshire from its call (on three occasions). The other four records of field signs are of droppings. Of the 10 records of dead animals, nine are of casual finds and one is of an animal killed by a domestic cat.

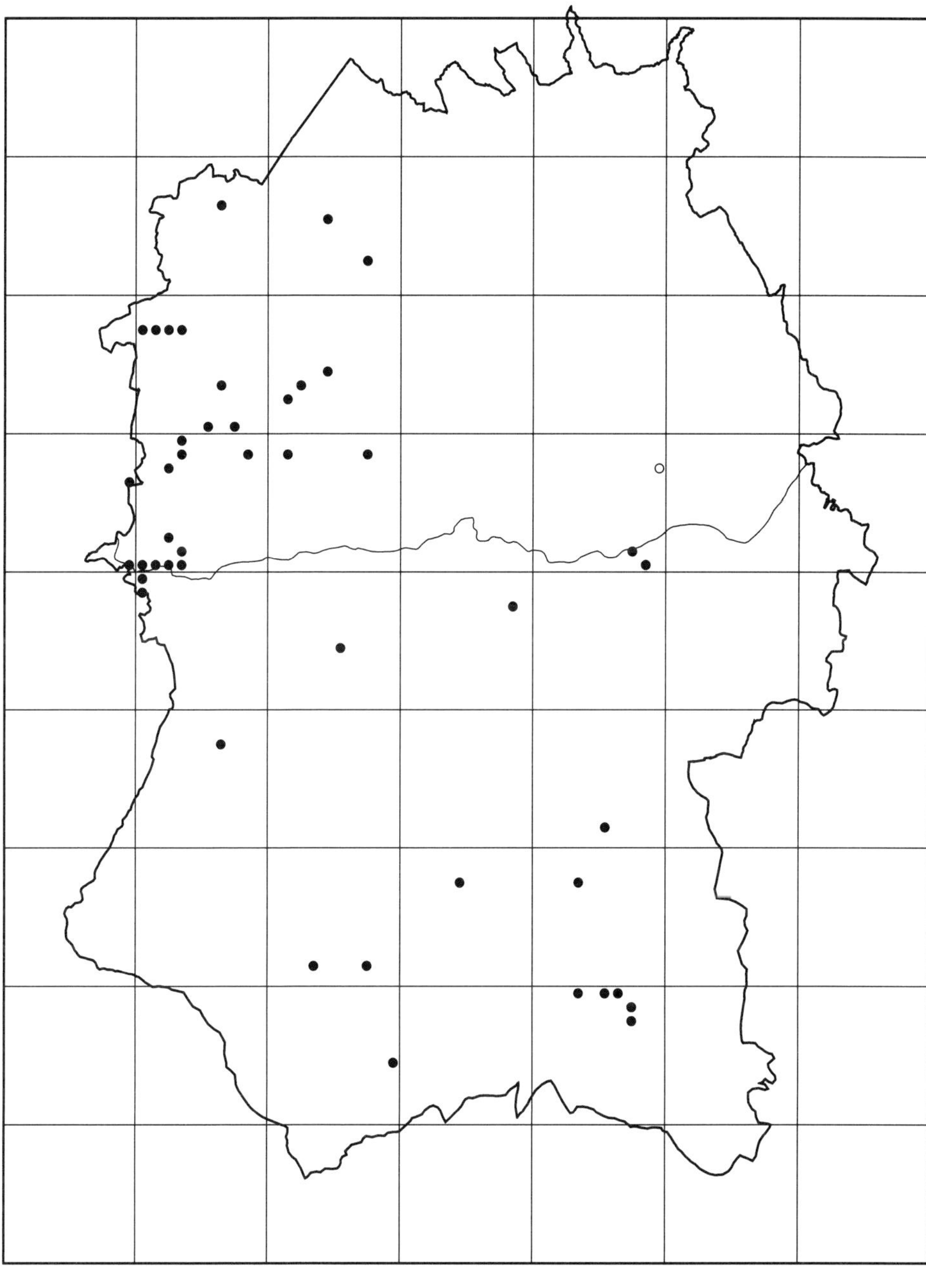

Fig. 12.1. Daubenton's bat: distribution

Order Chiroptera — Family Vespertilionidae

Serotine — *Eptesicus serotinus*

A large, dark brown bat with broad wings. Head and body 58-80 mm; wingspan 320-380 mm.

Handbook, 112-116; *Atlas,* 52-53; *Review,* 25-26; *Red Data Book,* 17-23.

Status

Native. Vulnerable.

The total pre-breeding population in Britain is estimated at 15,000. Status in recent historic times may have changed little; the population is now probably in decline.

Nationally

Confined mainly to south and south-east England, often associated with open flat country with hedgerows, woodland edge and parkland.

Wiltshire

First published record from Ludgershall (*SDNHS Report* 1958).

A general appeal for records was made in 1982 and, in 1985, a survey of churches was launched, but no detailed account of the bats has been published.

80 records of which 74 have been used in this analysis. 78% of the records are of field signs and of these, 58 in total, all but one are of droppings. Of the remaining records, 14% are of live and 8% are of dead animals. Widespread distribution in Wiltshire (fig. 13.1), mainly associated with roosting sites in domestic buildings (36) and churches (24) where searches for droppings have been successful. Probably under-recorded. Of the six records of dead animals, five are of casual finds and one is of an animal shot.

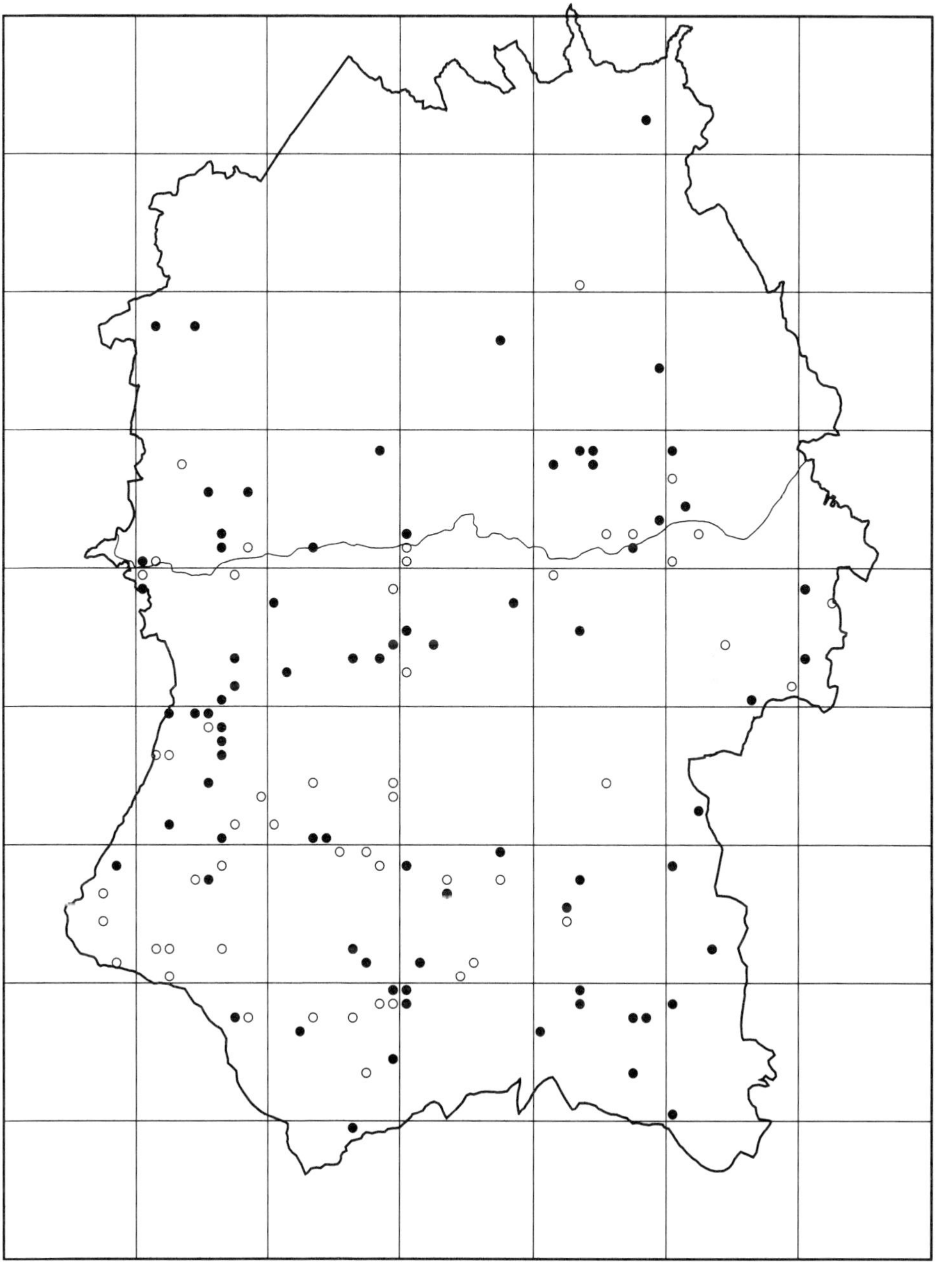

Fig. 13.1. Serotine: distribution

Order Chiroptera — Family Vespertilionidae

Noctule — *Nyctalus noctula*

A large bat with golden, ginger or reddish fur on the upper parts of the body, paler on the under parts, rounded ears and a short, mushroom-shaped tragus (a lobe developed from the lower rim of the ear). The only British species with which the Noctule can be confused is the smaller Leisler's bat but this has not been recorded in Wiltshire. Head and body 70-82 mm; wingspan 330-450 mm.

Handbook, 117-121; *Atlas,* 54-55; *Review,* 26-27; *Red Data Book,* 17-23.

Status

Native. Status said to be vulnerable (*Atlas*) and generally uncommon (*Review*).

The total pre-breeding population in Britain is estimated at 30,000-50,000. Evidence of a decline in recent decades and this is likely to continue as a result of further habitat fragmentation.

Nationally

Widely distributed in England and Wales, usually associated with woodland habitats.

Wiltshire

A general appeal for records was made in 1982 and, in 1985, a survey of churches was launched, but no detailed account of the bats has been published.

78 records of which 45 have been used in this analysis. 76% of the records are of live animals and 24% are of field signs. Wide but scattered distribution in Wiltshire (fig. 14.1). Associated with a range of habitats and roosting sites – gardens (10) wetlands (6), quarries (6), churches (2) with one record each from downland and parkland. Probably under-recorded. One of three bat species to be successfully identified in Wiltshire from its call (on nine occasions). The other two records of field signs are of droppings. No information on mortality.

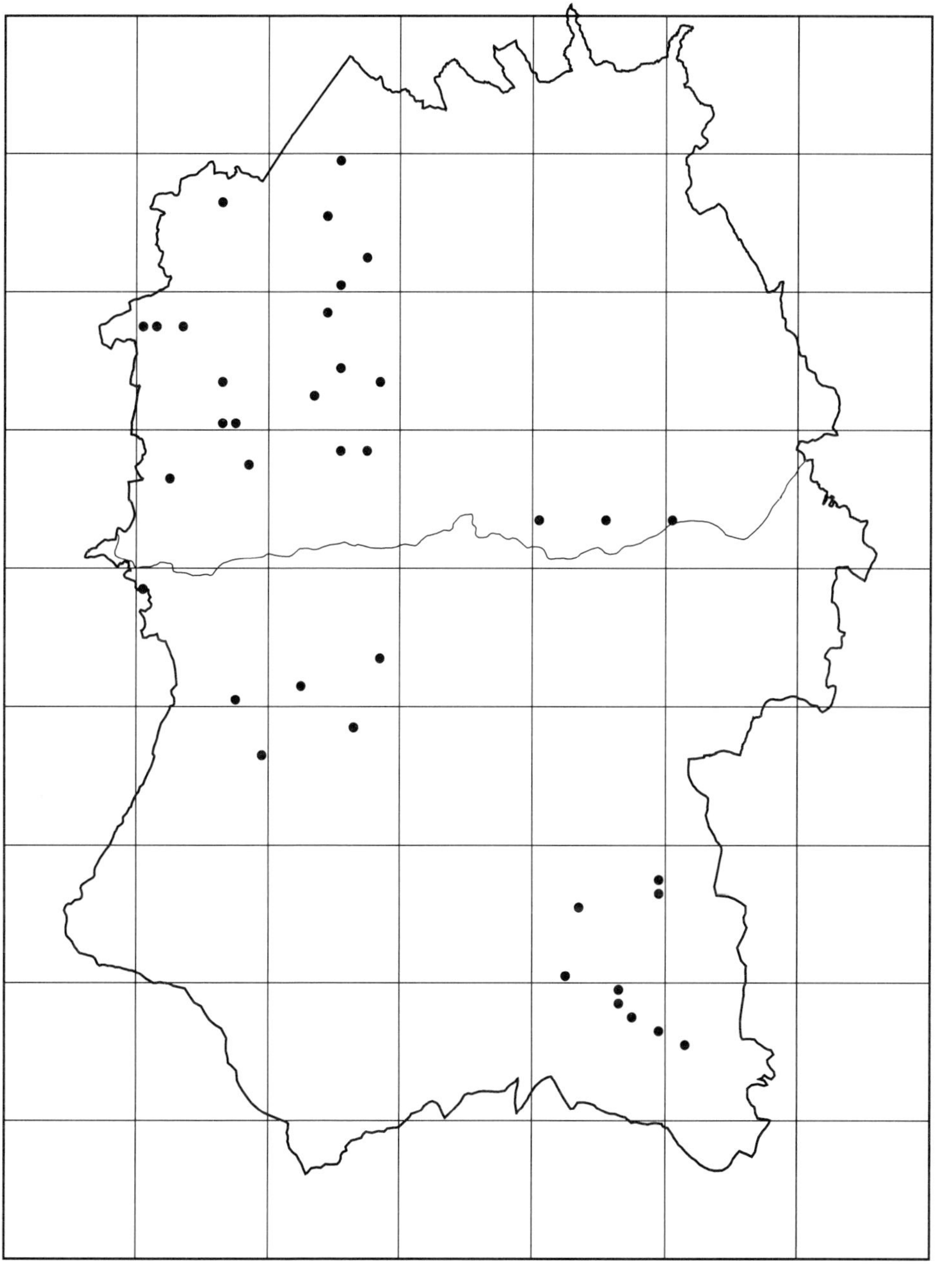

Fig. 14.1. Noctule: distribution

Order Chiroptera — Family Vespertilionidae

Pipistrelle — *Pipistrellus pipistrellus*

The smallest British bat with short ears and a short, curved, blunt tragus (a lobe developed from the lower rim of the ear). Head and body 35-45 mm; wingspan 190-250 mm.

Handbook, 124-128; *Atlas,* 58-59; *Review,* 27-29; *Red Data Book,* 17-23.

Status

Native. Views about status vary from vulnerable (*Atlas*) to common in most areas (*Review*).

The total pre-breeding population in Britain is estimated at 2,000,000. Population thought to have declined in recent decades.

Nationally

The commonest and most widespread British bat. Widely distributed in England, Wales and Scotland. In summer it frequently roosts in houses, modern as well as old, and for this reason it is the species most encountered by the public.

Wiltshire

First recorded in 1880 at Hilmarton (Goddard 1942). Said by Meyrick (1896) to be "very common everywhere" and by Hony (1917) to be "generally distributed".

A general appeal for records was made in 1982 and, in 1985, a survey of churches was launched, but no detailed account of the bats has been published.

493 records of which 457 have been used in this analysis. 59% of the records are of live animals, 27% are of field signs and 14% are of dead animals. Widespread distribution in Wiltshire (fig. 15.1) Associated with a range of habitats, particularly domestic buildings (150 out of 309 records). Other habitats recorded are churches (66), gardens (52), wetlands (16), roadbanks and railbanks (11), farmland (7), and woodland (7). Despite the wide coverage of the county in terms of records, the pipistrelle is probably under-recorded. The 125 records of field signs are of droppings, usually at roost sites. Of the 61 records of dead animals, 51 are of casual finds, eight are of animals killed by domestic cats, one is of an animal dead on a road, and one is of an animal taken by a kestrel.

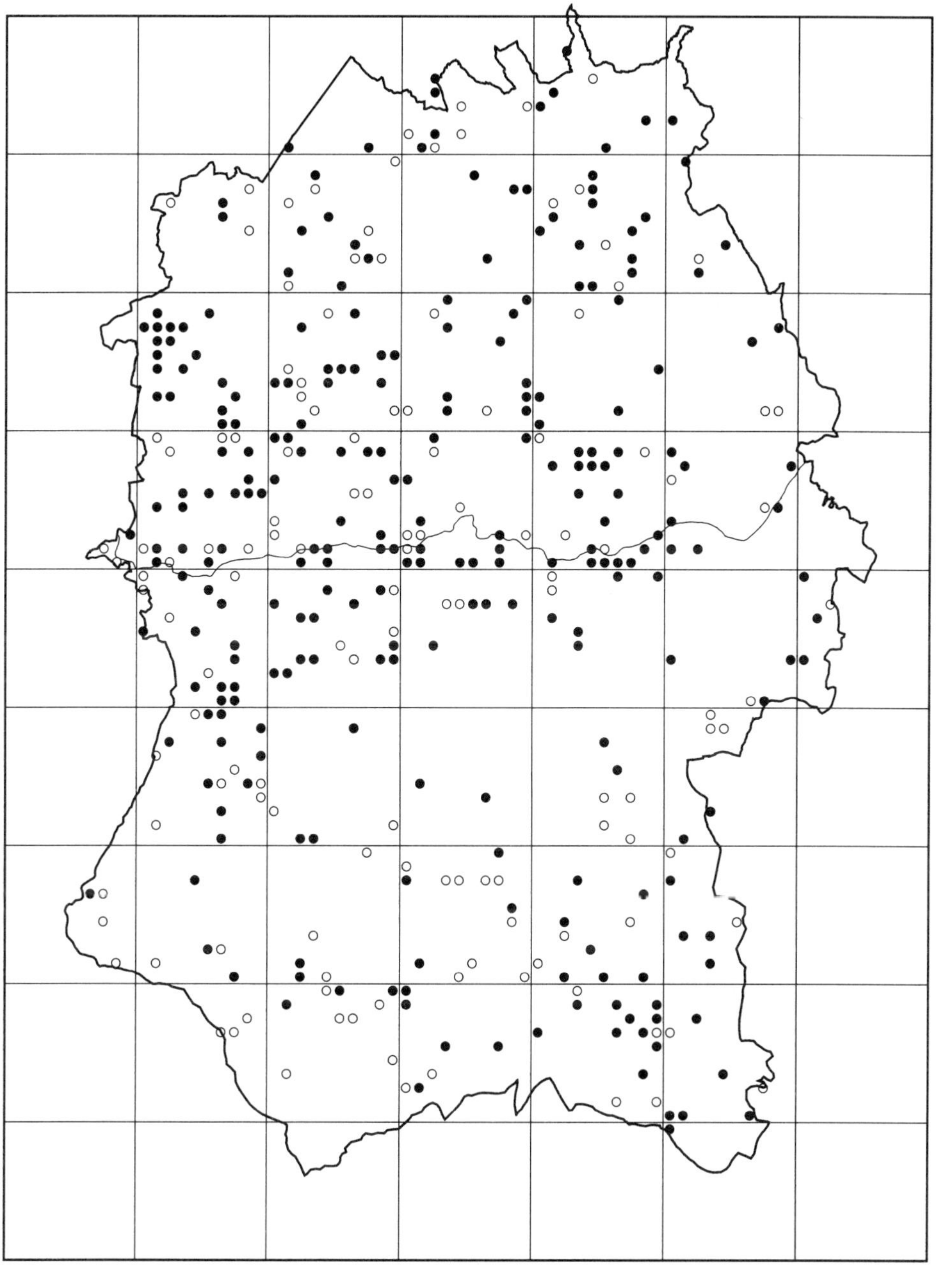

Fig. 15.1. Pipistrelle: distribution

Order Chiroptera — Family Vespertilionidae

Barbastelle — *Barbastella barbastellus*

A small bat with distinctive broad, short ears which meet at the centre of the forehead. Head and body 40-52 mm; wingspan 245-280 mm.

Handbook, 128-131; *Atlas,* 60-61; *Review,* 30; *Red Data Book,* 17-23.

Status

Native. Rare.

The total pre-breeding population in Britain is estimated at 5,000. Status in historic times and present population trends unknown.

Nationally

A scattered and sparse distribution in England, mainly in the south.

Wiltshire

The first record is of a specimen knocked down with a stick in broad daylight in Salisbury in 1869 (*Zoologist* 1869, p.1558).

A general appeal for records was made in 1982 and, in 1985, a survey of churches was launched, but no detailed account of the bats has been published.

Six records of which four have been used in this analysis. Of the four records, two are of live and two of dead animals. All records are from the south of the county (fig. 16.1). Of the three records which include details of habitat, two are of quarries, and one is of woodland.

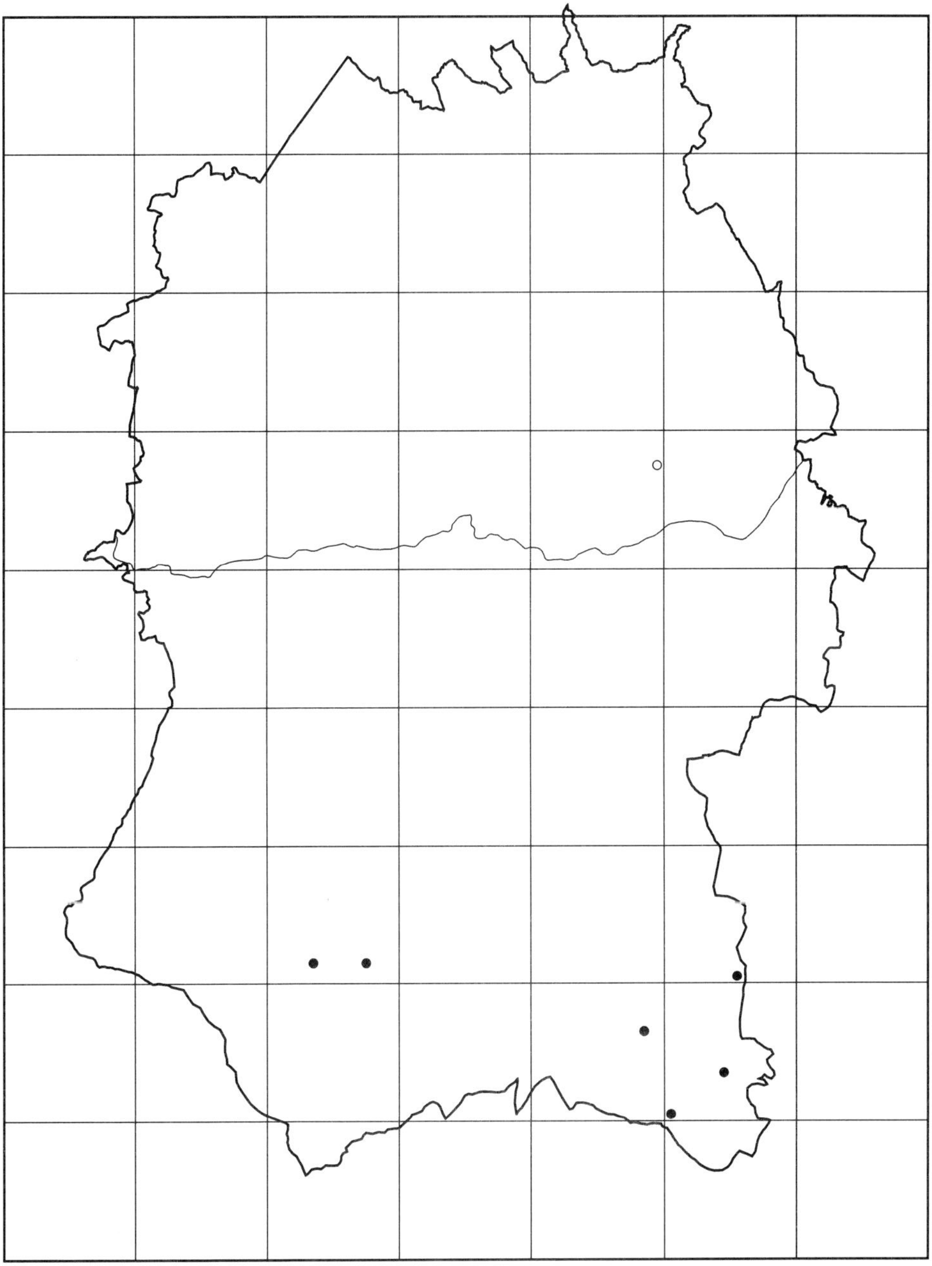

Fig. 16.1. Barbastelle: distribution

Order Chiroptera — Family Vespertilionidae
Brown long-eared bat — *Plecotus auritus*

A small to medium sized bat with distinctive long ears which meet above the forehead. Head and body 37-48 mm; wingspan 230-285 mm.

Handbook, 131-135; *Atlas,* 64-65; *Review,* 30-31; *Red Data Book,* 17-23.

Status

Native. Views about status vary from vulnerable (*Atlas*) to common (*Review*).

The total pre-breeding population in Britain is estimated at 150,000-200,000. There has been a long-term population decline which is likely to continue as the species is susceptible to timber treatment in buildings.

Nationally

Widespread distribution in England, Wales and Scotland and relatively common. It is often found roosting in houses.

Wiltshire

The first record is of six caught by hand by E.H. Goddard at Hilmarton in 1880 (Goddard 1942).

A general appeal for records was made in 1982 and, in 1985, a survey of churches was launched, but no detailed account of the bats has been published.

304 records of which 302 have been used in this analysis. 47% of the records are of field signs, 34% are of live and 19% are of dead animals. Widespread distribution in Wiltshire (fig. 17.1). Associated almost exclusively with roosting sites in domestic buildings and churches (107 and 54 respectively out of 167 habitat records). Despite the wide coverage of the county in terms of records, the brown long-eared bat is probably under-recorded. All 140 records of field signs are of droppings. Of the 58 records of dead animals, 48 are of casual finds, seven are of animals killed by domestic cats, one is of an animal dead on a road, one is of an animal drowned and one is of an animal poisoned.

A closely related species, the grey long-eared bat, found in parts of central Europe, is now known to have a range which extends into southern England. The grey long-eared bat has not been recorded in Wiltshire, although it is known from the neighbouring counties of Hampshire, Dorset and Somerset. For the purposes of this analysis, all long-eared bat records from Wiltshire have been assumed to be of the brown long-eared bat.

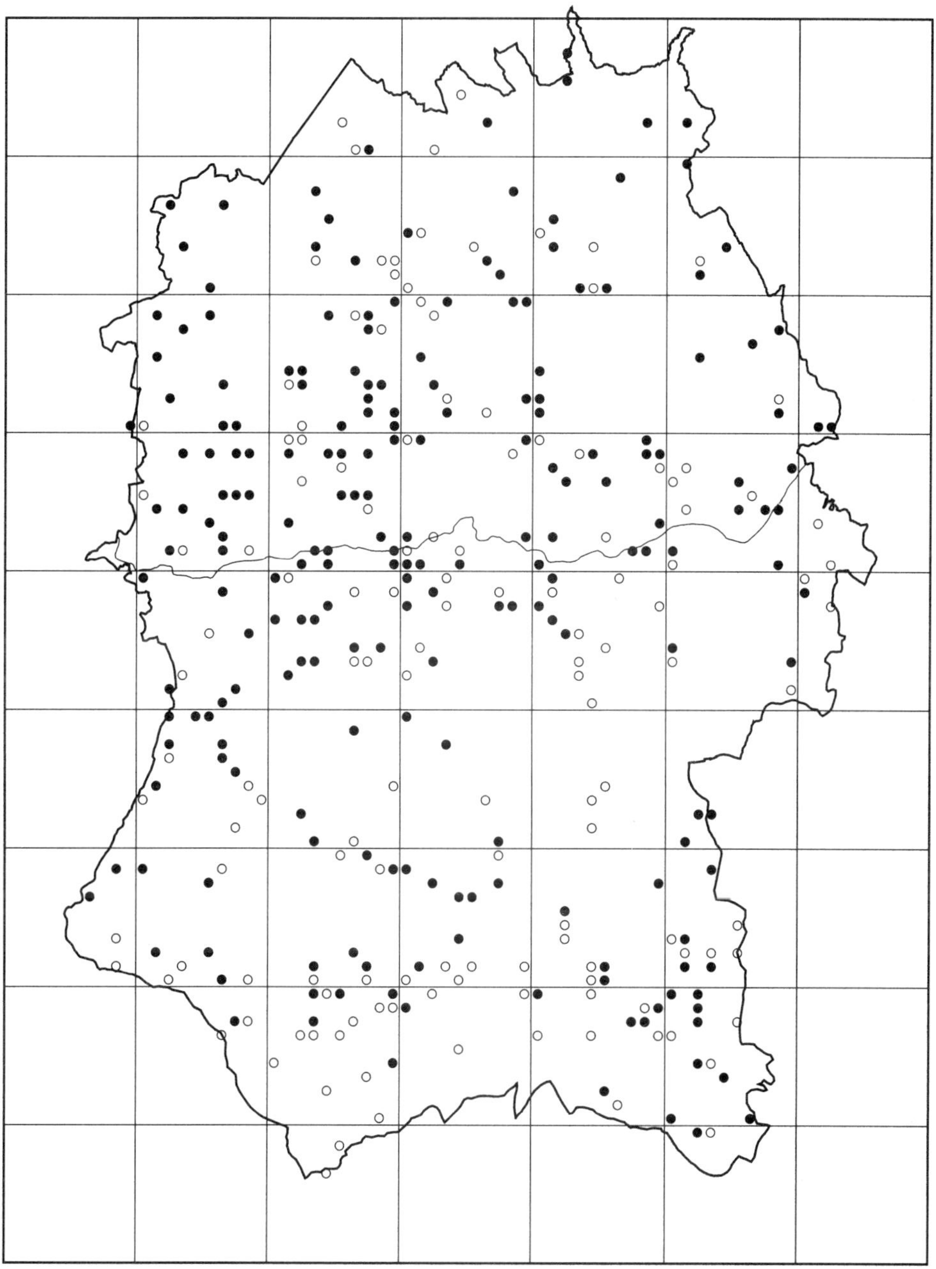

Fig. 17.1. Brown long-eared bat: distribution

Order Lagomorpha — Family Leporidae

Rabbit — *Oryctolagus cuniculus*

Easily recognisable medium sized mammal with distinctive long ears and a short, white tail. Can only be confused with the hares which are larger and have black tips to their ears. Head and body up to 400 mm.

Handbook, 146-154; *Atlas,* 68-69; *Review,* 32-34; *Game Heritage,* 22-25.

Status

Introduced. Common

The total pre-breeding population in Britain is estimated at 37,500,000.

Numbers are thought to be increasing although locally there may be considerable fluctuations in numbers.

Nationally

Virtually ubiquitous in England, Wales and Scotland. Particularly found in grassland and farmland habitats especially where there are associated woodland edge, hedgerow and marginal areas.

Wiltshire

First recorded as a warren animal by Aubrey (c.1660): "I come now to the warrens. That at Auburn [Aldbourne] is our famous coney-warren; and the conies there are the best, sweetest and fattest in the land."

No special survey of the rabbit has been published but for an account of the rabbit warren at Fyfield Down near Marlborough see King and Sheail (1970).

1686 records, of which 1482 have been used in this analysis. 67% of the records are of live animals (fig. 18.1). Widespread distribution (fig. 18.2); gaps in the record are probably due to lack of recording. Present in a wide range of habitats, particularly downland, farmland, road- and railbanks and woodland (fig. 18.3). Of the 105 records of field signs, 76 are of droppings. Other field signs recorded include burrows (14), prints (10), and a single incidence of remains in a bird pellet. Most recorded mortality due to road casualties (347 of 383 records). Only three records of mortality due to *myxoma*, and three records of animals shot. A moderate range of predators recorded with no species predominating: domestic cat (6), fox (4), stoat (4), domestic dog (1) and barn owl (1).

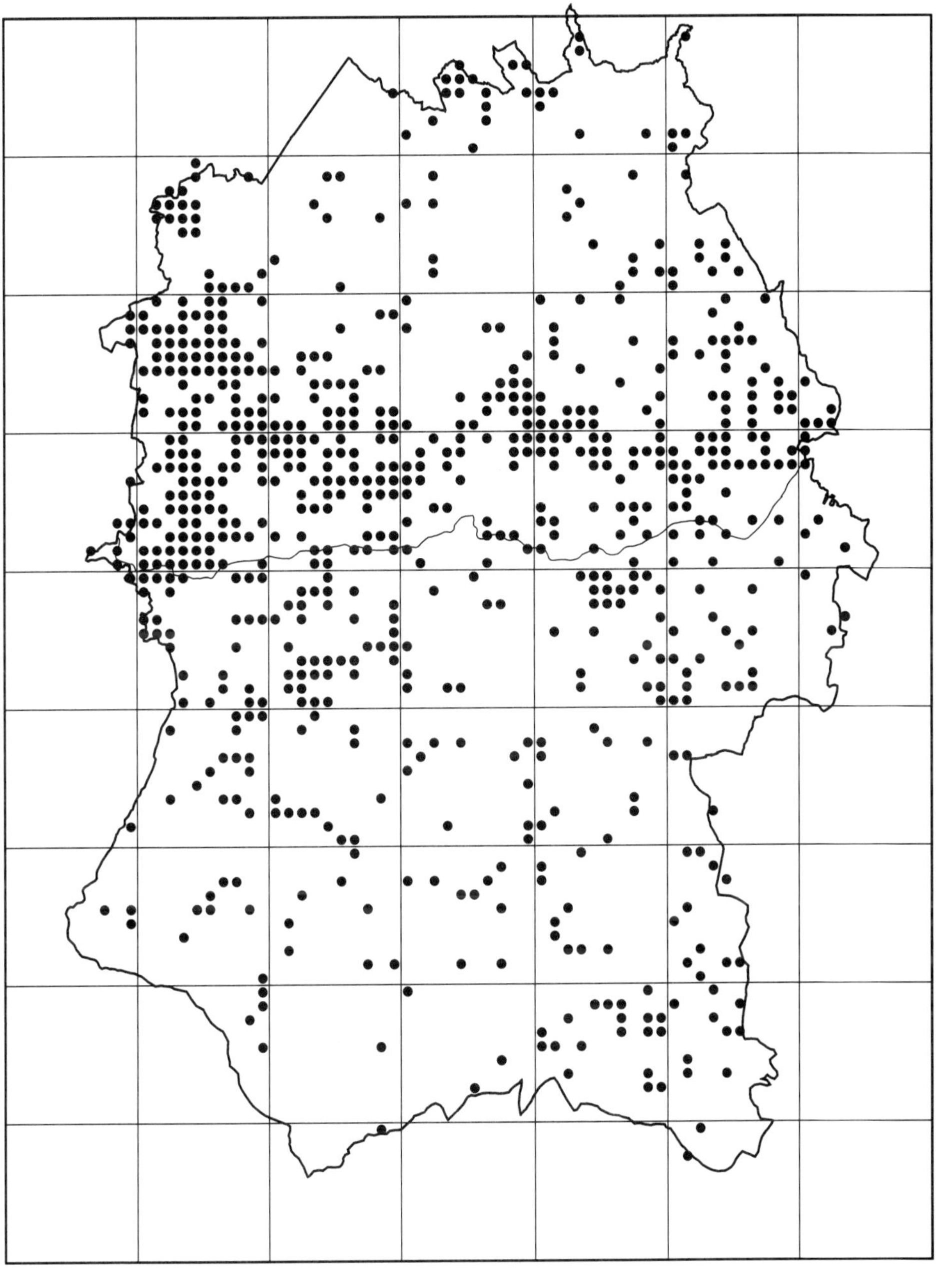

Fig. 18.2. Rabbit: distribution

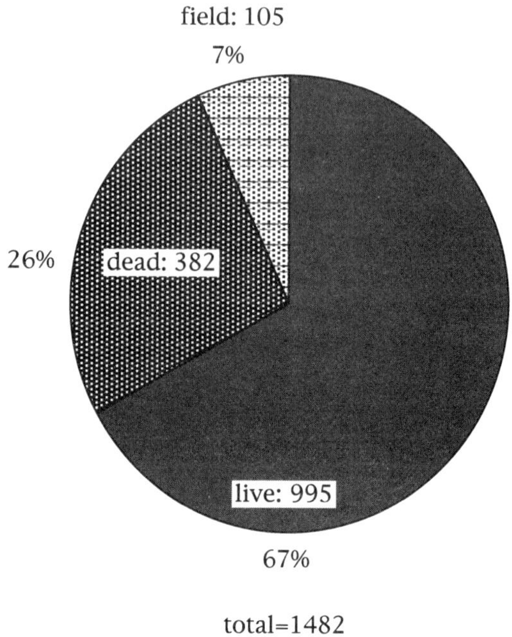

Fig. 18.1. Rabbit: number and nature of record

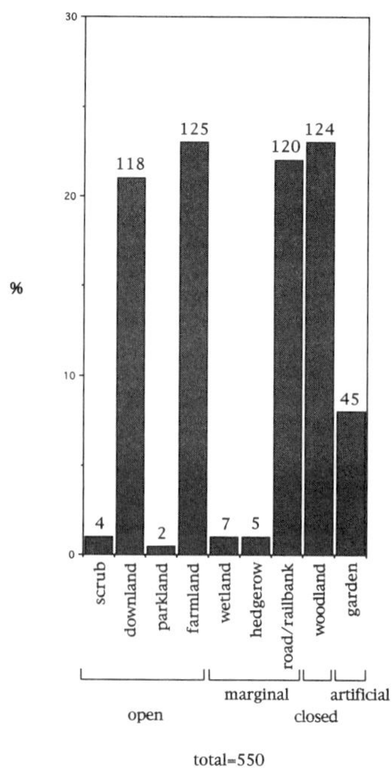

Fig. 18.3. Rabbit: habitats

Order Lagomorpha — Family Leporidae

Brown hare — *Lepus europaeus*

Easily recognisable medium sized mammal with distinctive long, black-tipped ears. Can only be confused with the rabbit and the mountain hare. It is larger than the former, which also lacks the black tips to the ears; the latter has not been recorded in Wiltshire. Head and body up to 595 mm.

Handbook, 154-161; *Atlas,* 70-71; *Review,* 34-37; *Red Data Book,* 24-28; *Game Heritage,* 26-29.

Status

Possibly introduced. Common.

The total pre-breeding population in Britain, before the onset of the main culling season, is estimated at 817,500. The population has undergone a substantial decline since the turn of the century and may now have stabilised at a low level. Locally there may be considerable fluctuations in numbers as a result of shooting and hunting and the influence of weather on breeding success.

Nationally

Widespread in England, Wales and Scotland but absent from some upland areas. Particularly found in open habitats.

Wiltshire

Aubrey (c.1660) writes of a "great warren for hares" near Wilton and of another at West Lavington made by the Earl of Abingdon. It may have been the belief in the seventeenth century that hares, like rabbits, could be managed in warrens.

No special survey of the brown hare has been published.

1130 records, of which 935 have been used in this analysis. 84% of the records are of live animals (fig. 19.1). Widespread distribution (fig. 19.2); gaps in the record are probably due to lack of recording. Present in a wide range of habitats (fig. 19.3), particularly downland and farmland. Of the 16 records of field signs, nine are of prints. Other field signs recorded are droppings (5) and a form. Most recorded mortality due to road casualties (102 of 141 records). Smaller numbers of records of animals 'killed' (18), shot (3), or 'killed by beagles' (2). The only recorded wild predator is the fox, a single incidence.

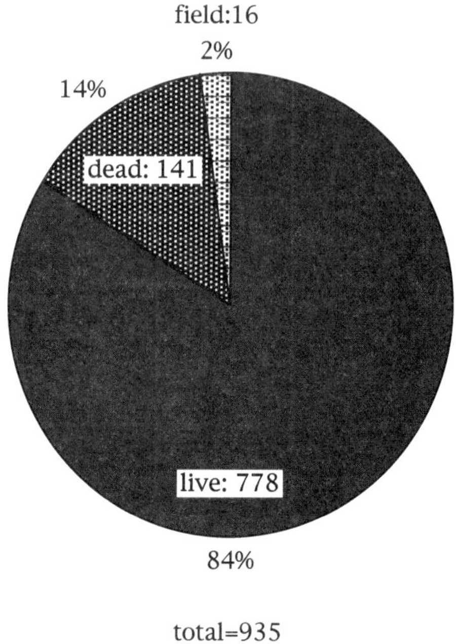

Fig. 19.1. Brown hare: number and nature of record

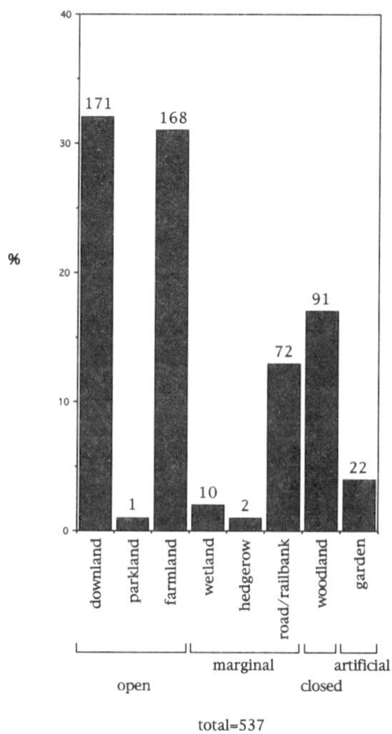

Fig. 19.3. Brown hare: habitats

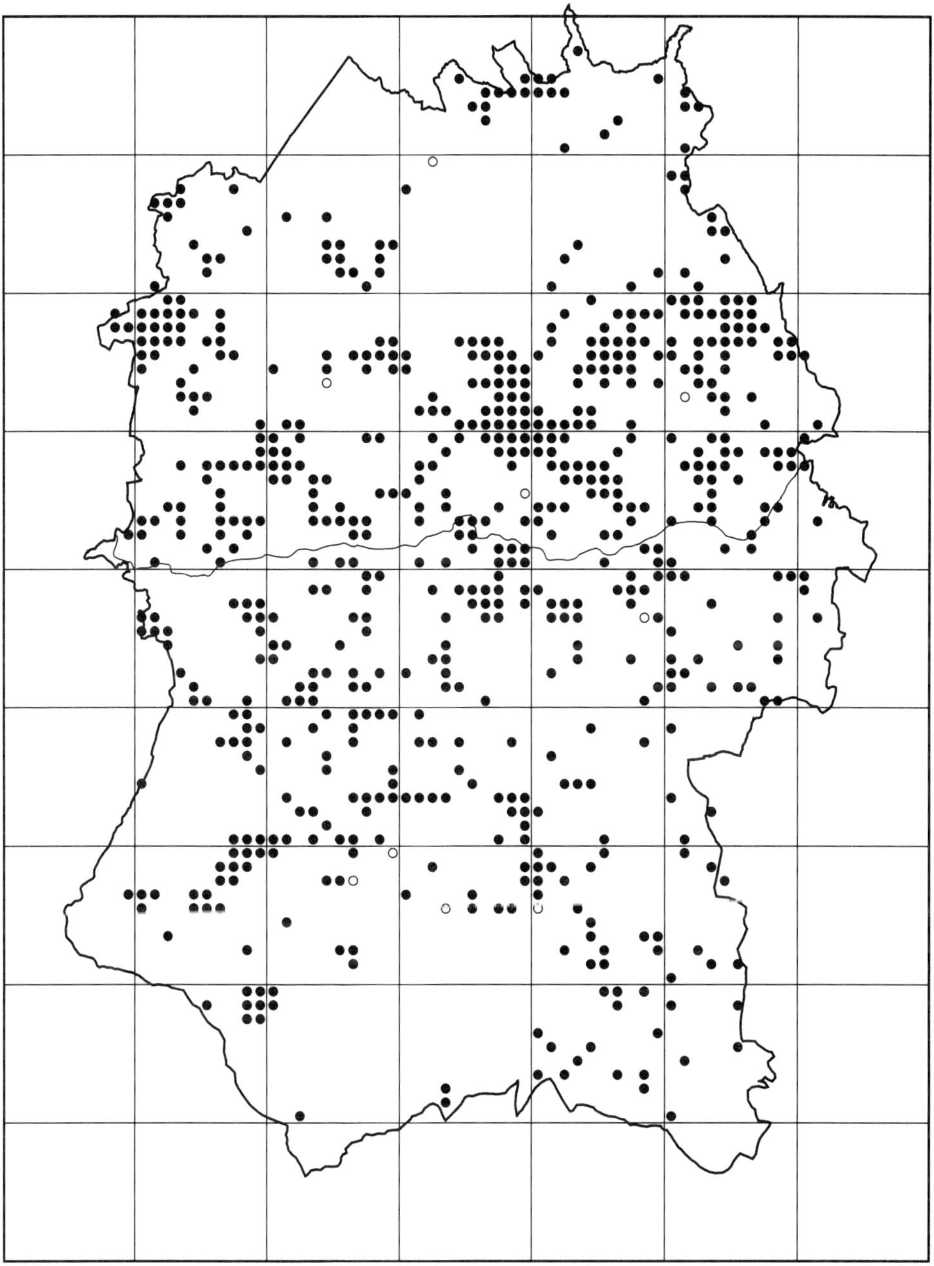

Fig. 19.2. Brown hare: distribution

Order Rodentia — Family Sciuridae

Red squirrel — *Sciurus vulgaris*

Easily recognisable medium sized mammal with distinctive bushy tail, conspicuous ear tufts (in winter) and reddish brown fur. Head and body up to 180-240 mm.

Handbook, 177-186; *Atlas,* 74-75; *Review,* 39-41; *Red Data Book,* 56-61.

Status

Native. Declining

The total pre-breeding population in Britain is estimated at 161,000. The population has undergone a substantial decline since the turn of the century in both range and numbers. Competitive exclusion by grey squirrels is thought to be the major cause of a continued decline.

Nationally

Common in Scotland, present in northern England, isolated populations in Wales, East Anglia and the Midlands. Thought to be absent from southern England apart from populations on the Isle of Wight and Brownsea Island. The red squirrel is a woodland species and in Britain is associated particularly with large tracts of coniferous woodland and smaller mixed woodlands and copses where the grey squirrel is absent.

Wiltshire

No special survey of the red squirrel has been published.

Not recorded since the 1960s although in the nineteenth century it was "more or less common in all woods" in the Marlborough district (Meyrick 1896). The decline appears to date from the late nineteenth century. The *Salisbury and Winchester Journal,* January 21st 1899, carried a letter protesting about the customary destruction of squirrels in the New Forest on Boxing Day. The writers describe the use of a "2 ft. lead instrument" for clubbing them to death. In 1910, Meyrick reported that the species was much less common in Savernake Forest and had become quite scarce *(MCNHS Report).* In 1917 Hony reported that it was "generally distributed, but much scarcer lately". Peirson (1923) claimed that the species had disappeared over a large area in north Wiltshire by 1913. In neighbouring Dorset, red squirrels were found throughout most of the county in 1945 but by 1951 had disappeared from all but a few parishes (Benham 1953).

The last records in Wiltshire, most from *SDNHS Report,* are:

1948 Corsham Park

1956 Whiteparish Common, Clarendon and East Grimstead

1957 Clarendon, East Grimstead and Porton

1965 Clarendon, Alderbury

Order Rodentia — Family Sciuridae

Grey squirrel — *Sciurus carolinensis*

Easily recognisable medium sized mammal with distinctive body shape and tail. Fur mainly grey although there can be some brown colouration. Larger than the red squirrel with no conspicuous ear tufts. Head and body 240-285 mm.

Handbook, 186-191; *Atlas,* 76-77; *Review,* 41-43; *Game Heritage,* 84-85.

Status

Introduced. Common.

The total pre-breeding population in Britain is estimated at 2,520,000. The species continues to extend its range and the population is increasing.

Nationally

Introduced in the later nineteenth and early twentieth centuries and now widespread in England, Wales and parts of Scotland but absent from some upland areas. Found in a wide range of habitats, particularly woodland, farmland with good hedgerows, urban and suburban parks and gardens.

Wiltshire

No special survey of the grey squirrel has been published.

First recorded in Wiltshire, "with regret", in 1929 in Savernake Forest (*MCNHS Report* 78). Recorded in the Dauntsey district in 1931 *(DFFL)*. Widespread and common by the 1950s. The *Western Gazette,* March 20th, 1953, reported that 13,350 grey squirrels had been "bagged" in Wiltshire that year.

1133 records, of which 956 have been used in this analysis. 86% of the records are of live animals (fig. 20.1). Widespread distribution (fig. 20.2); gaps in the record where suitable habitat occurs are probably due to lack of recording. Present in a wide range of habitats but predominantly parkland and woodland (fig. 20.3). Feeding signs (25) and nest/dreys (25) account for most of the 52 records of field signs. Most recorded mortality due to road casualties (72 of 96 records). Smaller numbers of records of casual finds (13) and animals trapped (5), shot (3), and drowned (2). The only recorded predation is by domestic cat, a single incidence.

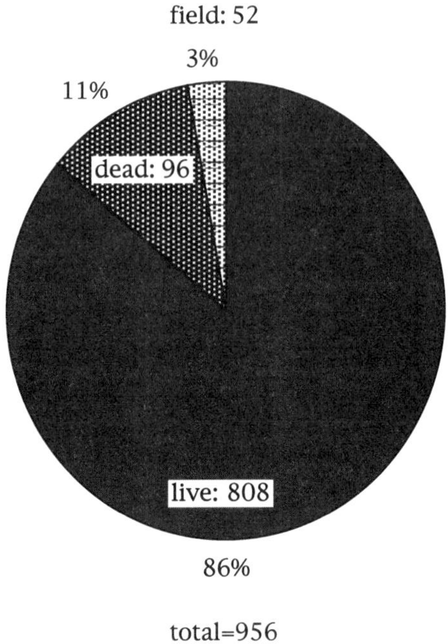

Fig. 20.1. Grey squirrel: number and nature of record

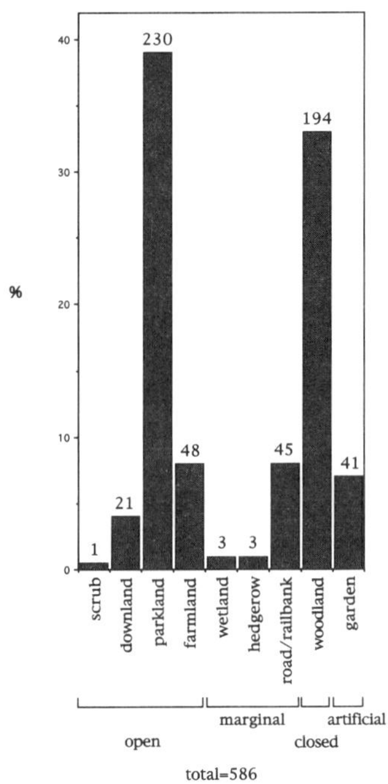

Fig. 20.3. Grey squirrel: habitats

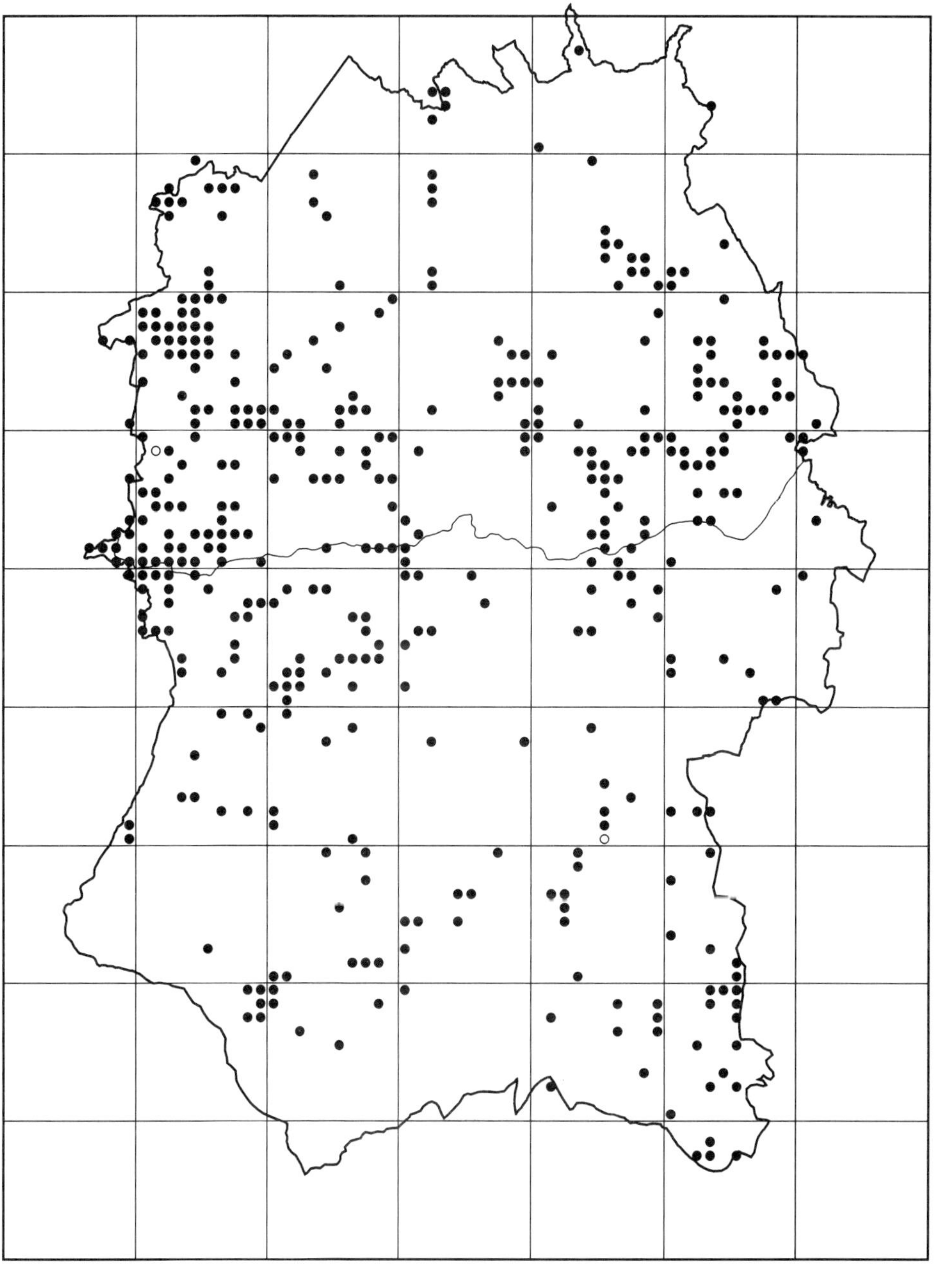

Fig. 20.2. Grey squirrel: distribution

Order Rodentia — Family Muridae

Bank vole — *Clethrionomys glareolus*

Small mammal with rounded head, blunt muzzle and plump body (fig. 21.1). Differs from the field vole in having a longer tail (about two-thirds the length of the body as opposed to about one-third) and chestnut-red fur on the upper side of the body. Head and body 90-110 mm. Foot and toe pads (fig. 21.2) and skull, jawbones and dentition (fig. 21.3) are diagnostic.

Handbook, 192-202; *Atlas,* 78-79; *Review,* 43-45.

Status

Native. Abundant.

The total pre-breeding population in Britain is estimated at 23,000,000 although this may fluctuate widely. Status in historical times and population trends are unknown.

Nationally

Widely distributed in England, Wales and Scotland. Found in a wide range of habitats, particularly woodland and marginal areas with good ground cover.

Wiltshire

Data up to 1983 published in *WAM* 78, 94-104.

238 records, of which 235 have been used in this analysis. 41% of the records are of live animals (fig. 21.4). Widespread distribution (fig. 21.5); gaps in the record where suitable habitat occurs are probably due to lack of recording. Present in a wide range of habitats but predominantly woodland, farmland and gardens (fig. 21.6). Remains in bird pellets account for over half the field sign records (fig. 21.7) but evidence from feeding (fig. 21.8) is also an important source of records. Wide range of mortality factors (fig. 21.9); the apparent predominance of predation by domestic cat and barn owl is probably a reflection of the data available from specific surveys ('what the cat brought in' (Ward 1981) and owl pellet workshops (see for example, Dillon 1977)).

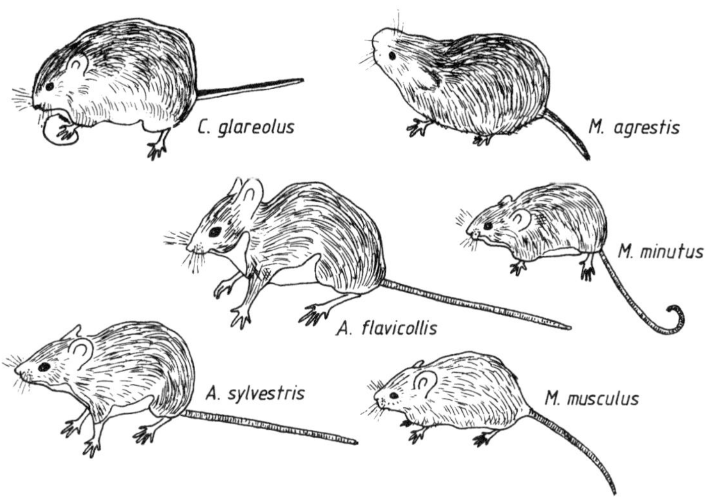

Fig. 21.1: Comparative sizes and proportions of small rodents, approximately x ¼

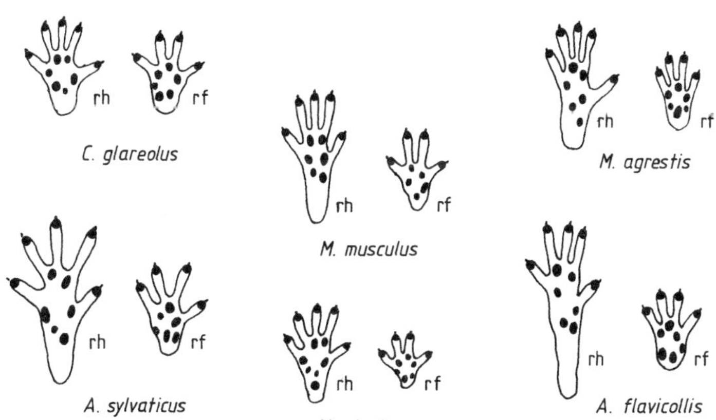

Fig. 21.2: Foot and toe pads of small rodents, approximately x 1

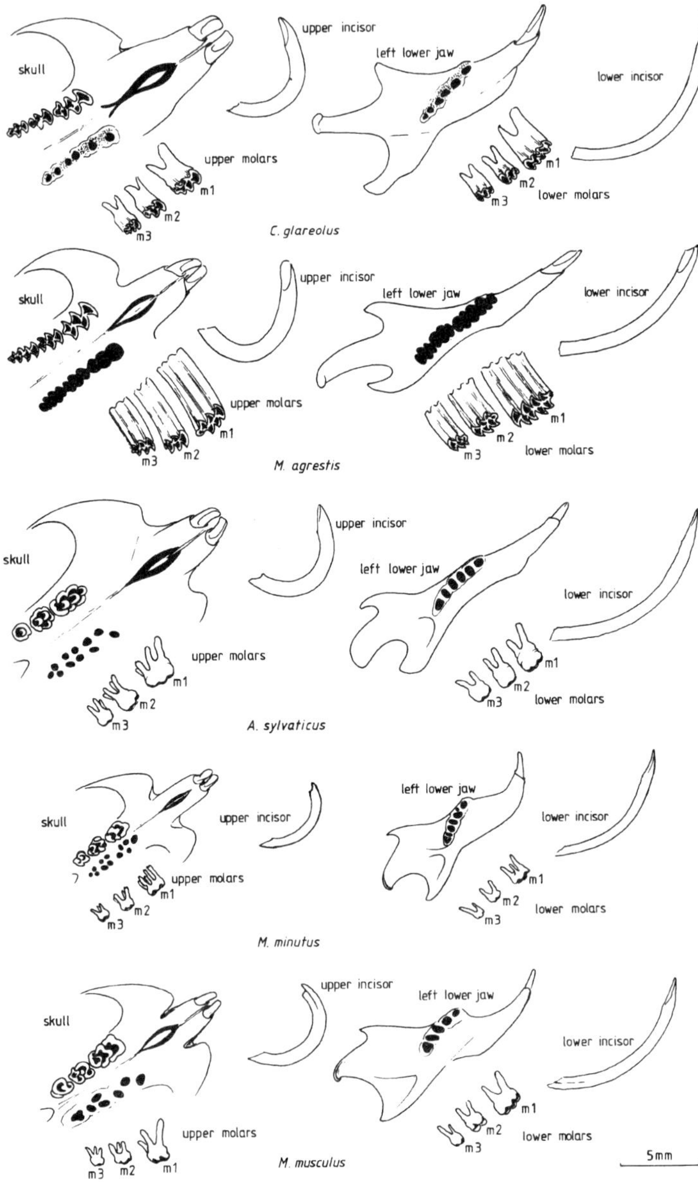

Fig. 21.3: Skulls, jawbones and dentition of small rodents, approximately x 2½

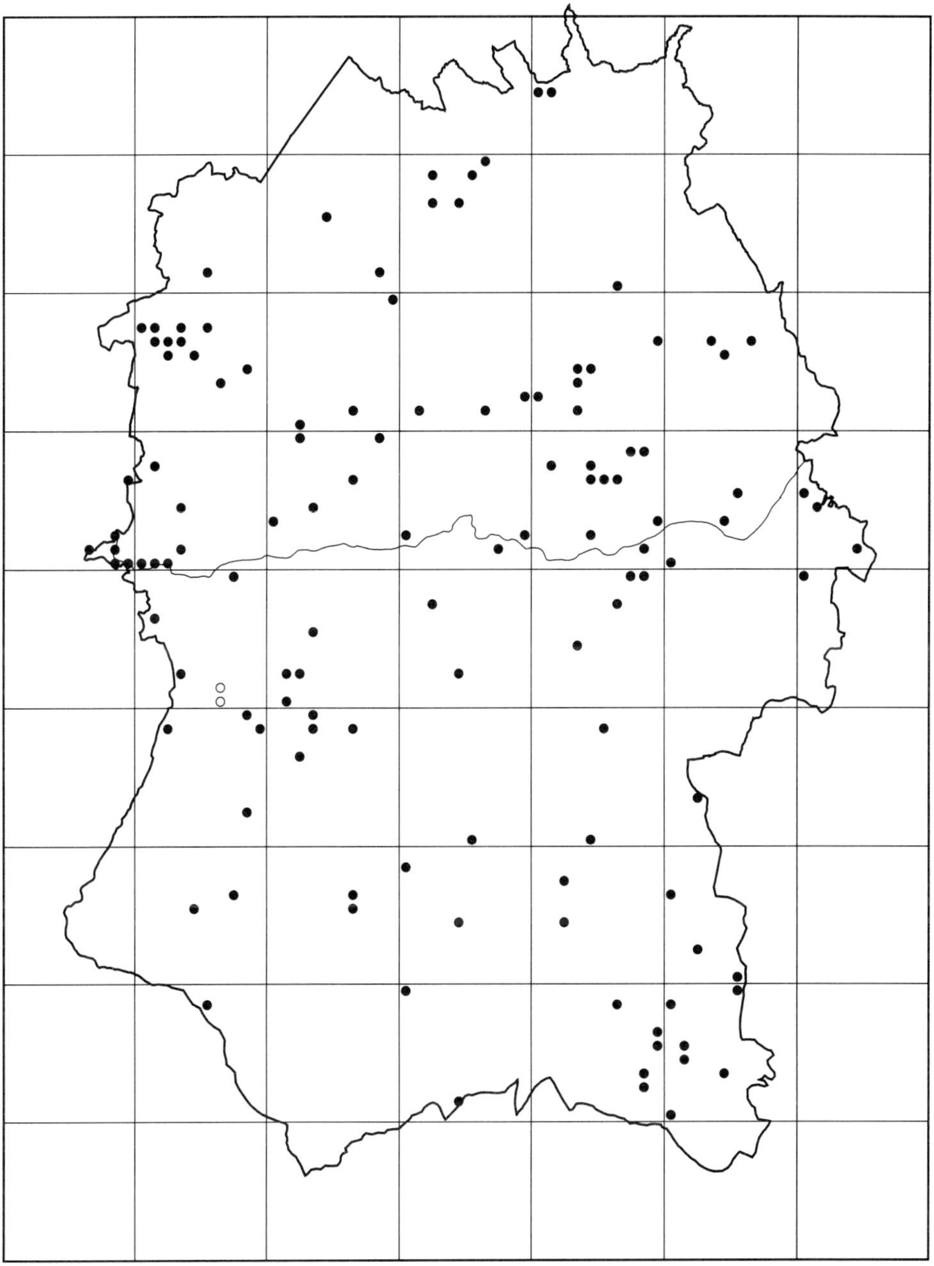

Fig. 21.5. Bank vole: distribution

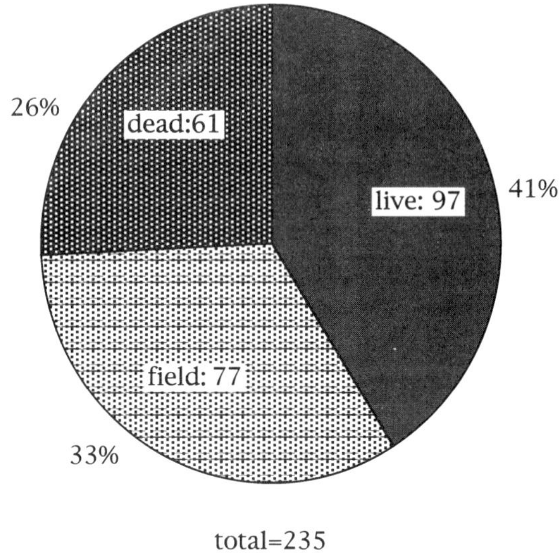

Fig. 21.4. Bank vole: number and nature of record

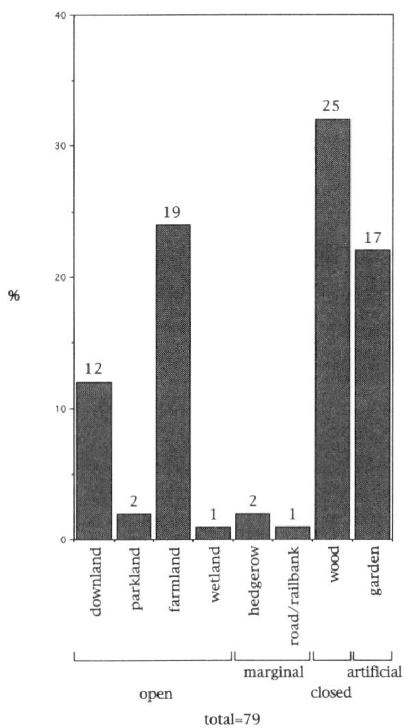

Fig. 21.6. Bank vole: habitats

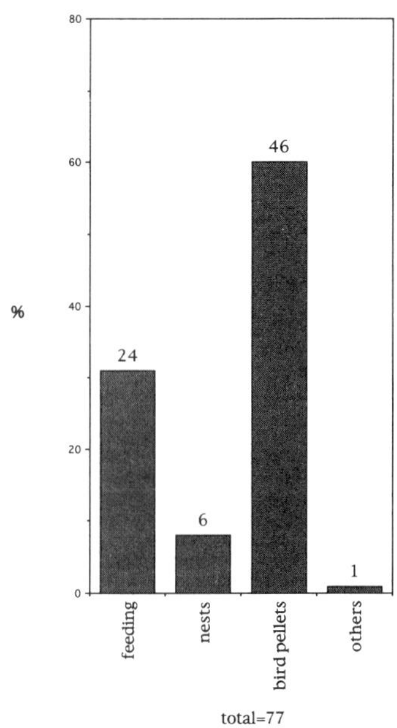

Fig. 21.7. Bank vole: field signs

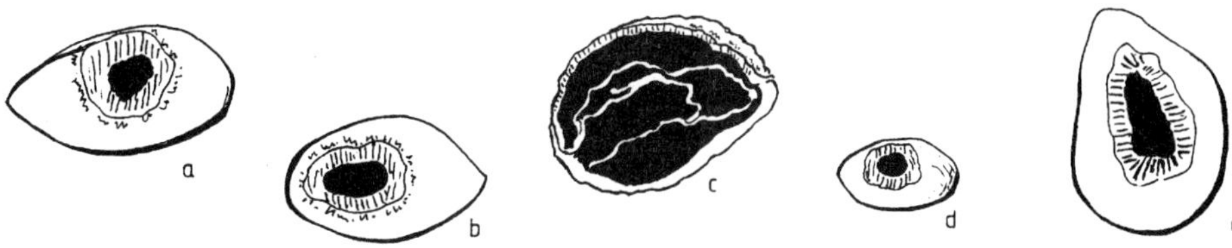

a) plum b) hazel c) walnut opened by *A. sylvaticus*. d) damson e) filbert opened by *C. glareolus*.

Fig. 21.8. Nuts and fruit stones opened by small rodents x 1

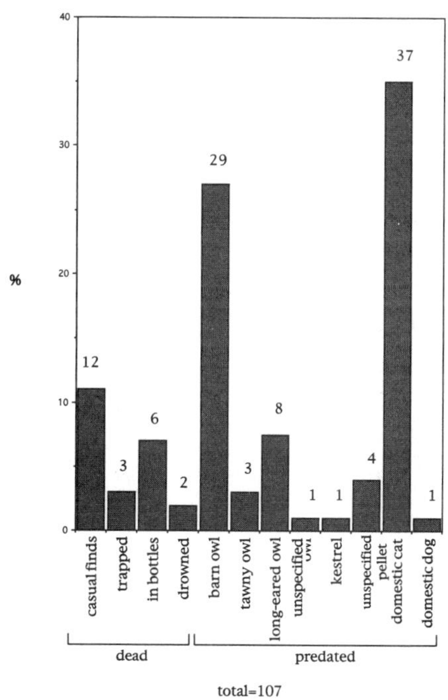

Fig. 21.9. Bank vole: mortality

Order Rodentia — Family Muridae
Field vole — *Microtus agrestis*

Also known as the short-tailed vole. Small mammal with rounded head, blunt muzzle and plump body (fig. 21.1). Differs from the bank vole in having a shorter tail (about one-third the length of the body as opposed to about two-thirds) and yellow-brown fur. Head and body 90-115 mm. Foot and toe pads (fig. 21.2) and skull, jawbones and dentition (fig. 21.3) are diagnostic.

Handbook, 203-208; *Atlas,* 80-81; *Review,* 45-48.

Status

Native. Abundant.

The total pre-breeding population in Britain is estimated at 75,000,000. The population has undergone a decline since the turn of the century as a result of habitat loss.

Nationally

Widely distributed in England, Wales and Scotland. Found in a wide range of habitats, particularly open grassland.

Wiltshire

Data up to 1983 published in *WAM* 78, 94-104.

270 records, of which 244 have been used in this analysis. 25% of the records are of live animals (fig. 22.1). Widespread distribution (fig. 22.2); gaps in the record where suitable habitat occurs are probably due to lack of recording. Present in a wide range of habitats but predominantly downland, farmland and gardens (fig. 22.3). Remains in bird pellets account for over three quarters of field sign records (fig. 22.4). Very wide range of mortality factors (fig. 22.5). The mammal with the largest range of predators recorded in Wiltshire; the apparent predominance of predation by domestic cat and barn owl is probably a reflection of the data available from specific surveys ('what the cat brought in' (Ward 1981) and owl pellet workshops (see for example, Dillon, 1977)).

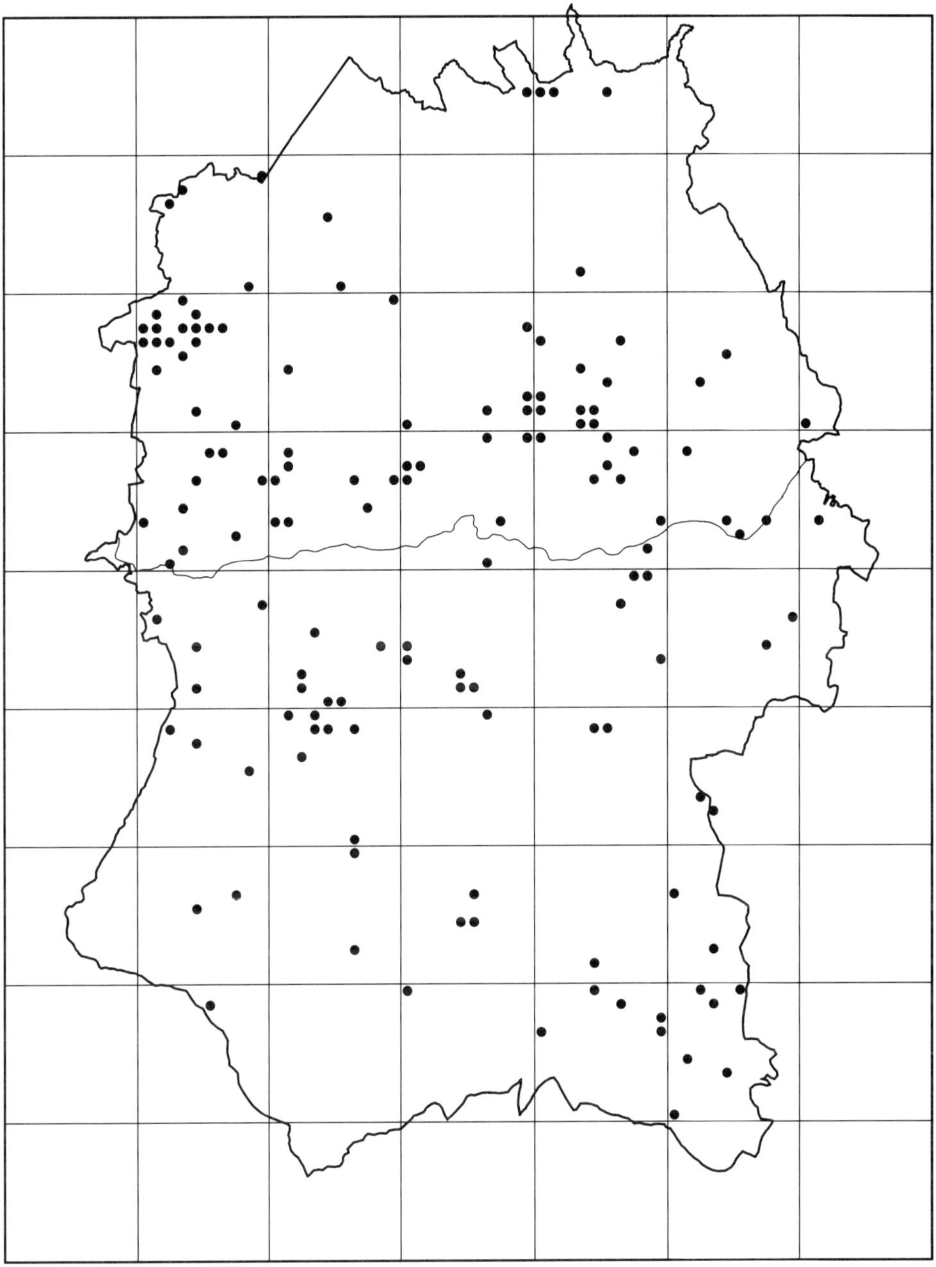

Fig. 22.2. Field vole: distribution

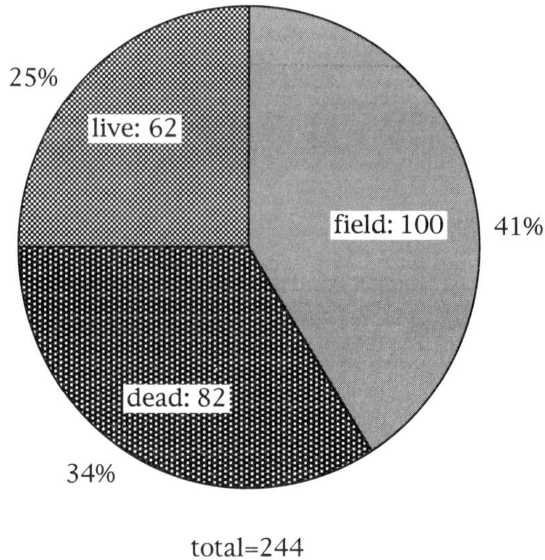

Fig. 22.1. Field vole: number and nature of record

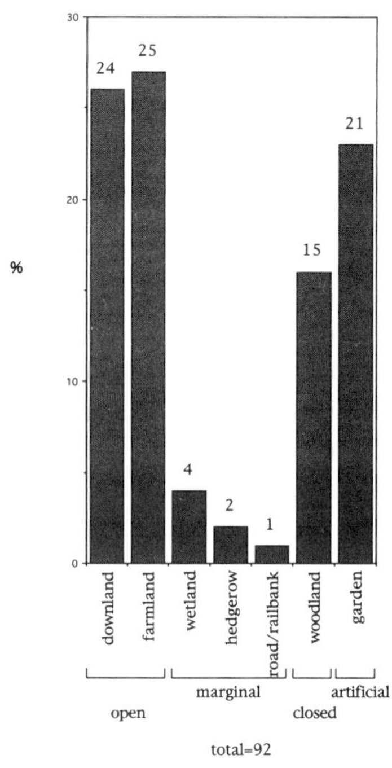

Fig. 22.3. Field vole: habitats

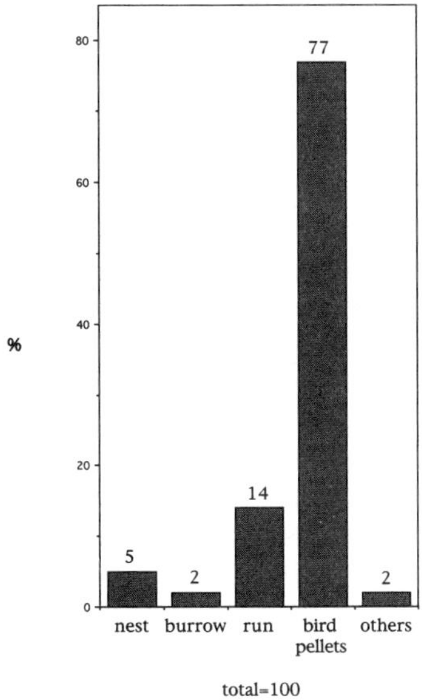

Fig. 22.4. Field vole: field signs

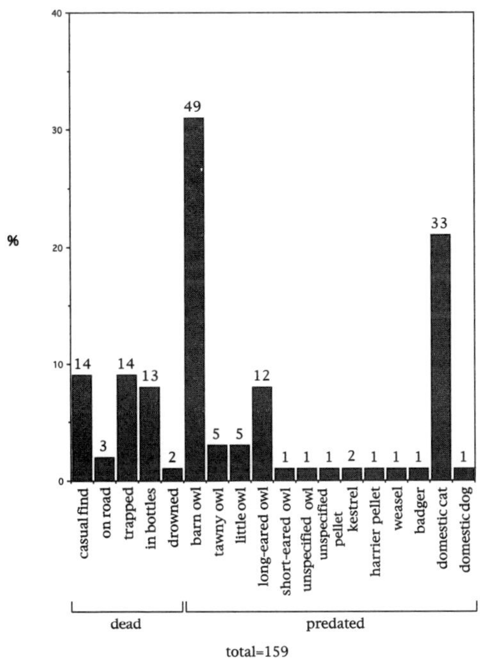

Fig. 22.5. Field vole: mortality

Order Rodentia — Family Muridae

Water vole — *Arvicola terrestris*

Small-medium sized mammal. The largest British vole with blunt rounded muzzle, small eyes and ears and a dense, shaggy fur (fig. 23.1). Can be confused with the common rat from which it may be distinguished by its facial features, short tail and foot and toe pads (fig. 23.2). Head and body 120-245 mm. Jaw bones and dentition are diagnostic (fig. 23.3).

Handbook, 212-218; *Atlas*, 84-85; *Review*, 49-51; *Red Data Book*, 51-55.

Status

Native. Common but declining.

The total pre-breeding population in Britain is estimated at 1,169,000. The population has undergone a substantial decline since the turn of the century. Continued decline is attributed to habitat loss, the use of pesticides and predation by mink.

Nationally

Widely distributed in England, Wales and parts of Scotland but absent from some upland areas. Associated mainly with a wide range of wetland habitats, particularly where there is dense bankside vegetation.

Wiltshire

Special appeal for records made 1980-81; data up to 1982 published in *WAM* 77, 123-137.

454 records, of which 354 have been used in this analysis. 90% of the records are of live animals (fig. 23.4). Widespread distribution, particularly associated with the major waterways of the county (fig. 23.5). Wetland accounts for over three-quarters of the records (fig. 23.6). A range of field signs are recorded (fig. 23.7). Of the 12 recorded deaths 2 each were recorded as casual finds or as road deaths, 1 each trapped or caught by barn owl, stoat and mink, and 4 predated by domestic cat.

Fig. 23.1: The water vole, approximately x ½

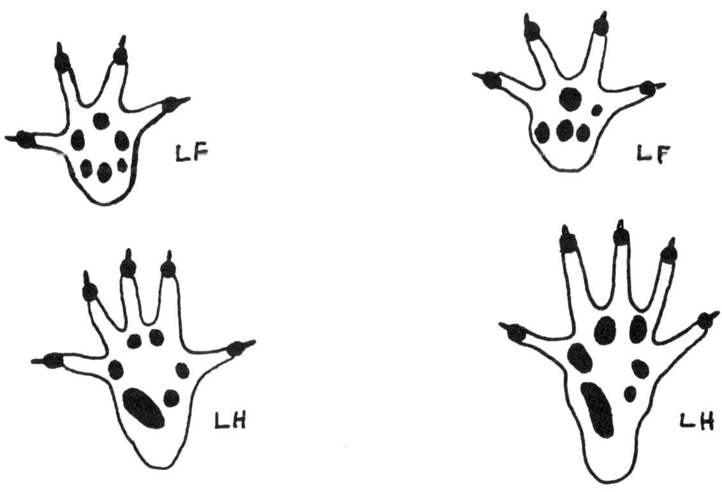

Fig. 23.2: Foot and toe pads of water vole (left) and common rat (right), approximately x 1

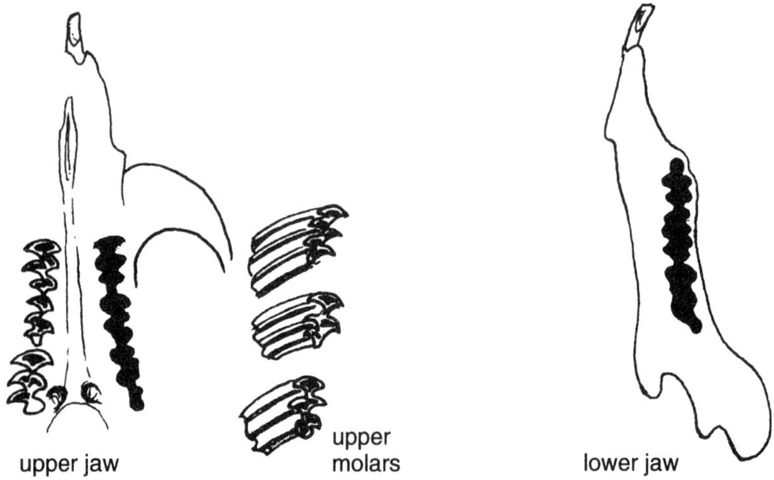

Fig. 23.3: Jawbones and dentition of water vole, approximately x2

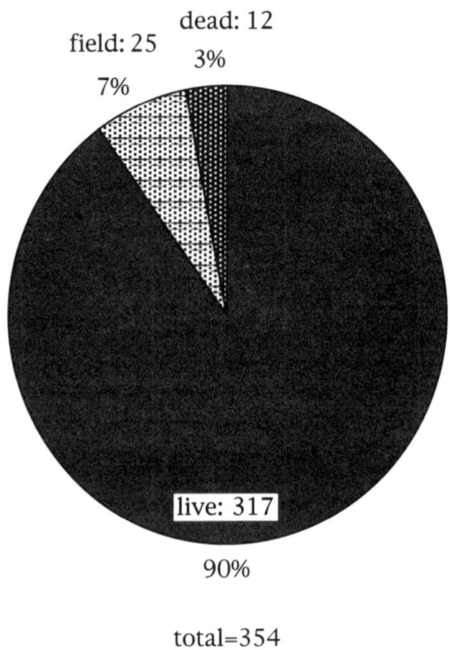

Fig. 23.4 Water vole: number and nature of record

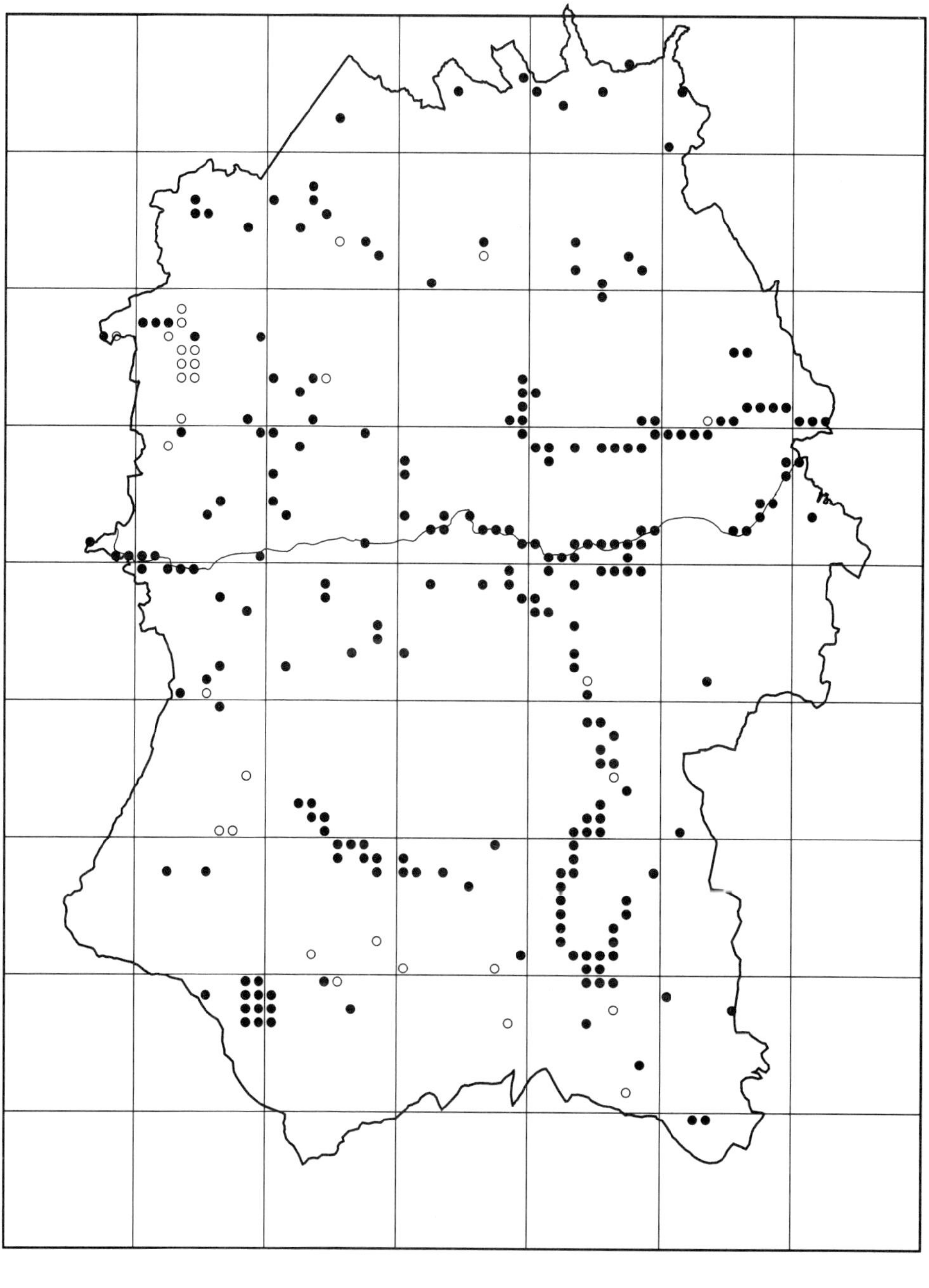

Fig. 23.5. Water vole: distribution

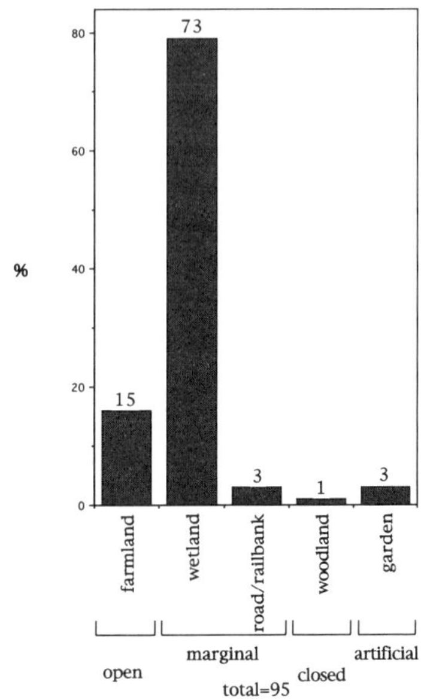

Fig. 23.6. Water vole: habitats

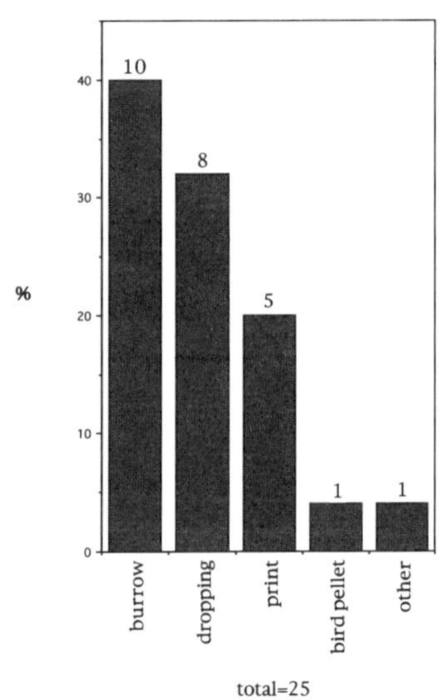

Fig. 23.7. Water vole: field signs

Order Rodentia — Family Murinae

Wood mouse — *Apodemus sylvaticus*

Also known as the long-tailed field mouse. Small mammal with a pointed muzzle, large ears and large black eyes (fig. 21.1). The fur is brown on the upper side of the body and pale grey on the underside. Can be confused with the yellow-necked mouse which is slightly larger and has a yellowish chest patch. Head and body 85-105 mm. Foot and toe pads (fig. 21.2) and skull, jawbones and dentition (fig. 21.3) are diagnostic.

Handbook, 220-229; *Atlas*, 86-87; *Review*, 51-52.

Status

Native. Abundant.

The total pre-breeding population in Britain is estimated at 38,000,000 although this may fluctuate widely. Status in historical times and population trends are unknown.

Nationally

Widely distributed in England, Wales and Scotland. Found in a wide range of habitats, particularly woodland and marginal areas with good ground cover.

Wiltshire

Data up to 1983 published in *WAM* 78, 94-104.

364 records, of which 302 have been used in this analysis. Similar proportions of records are of live animals, dead animals and field signs (fig. 24.1). Widespread distribution (fig. 24.2); gaps in the record where suitable habitat occurs are probably due to lack of recording. Present in a wide range of habitats but predominantly woodland and gardens (fig. 24.3). Comparatively large number of records from domestic buildings. Remains in bird pellets account for over half the records of field signs (fig. 24.4) but evidence from feeding (fig. 21.8) is an important source of rccords. Widc rangc of mortality factors, cspccially prcdators (fig. 24.5). Thc apparcnt predominance of predation by domestic cat and barn owl is probably a reflection of the data available from specific surveys ('what the cat brought in' (Ward 1981) and owl pellet workshops (see for example, Dillon, 1977)).

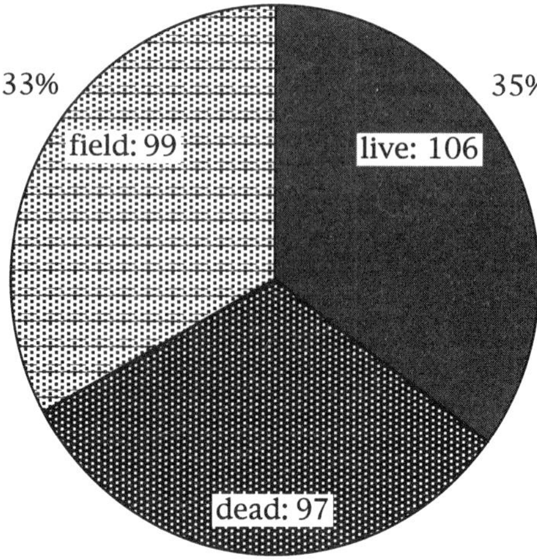

Fig. 24.1. Wood mouse: number and nature of record

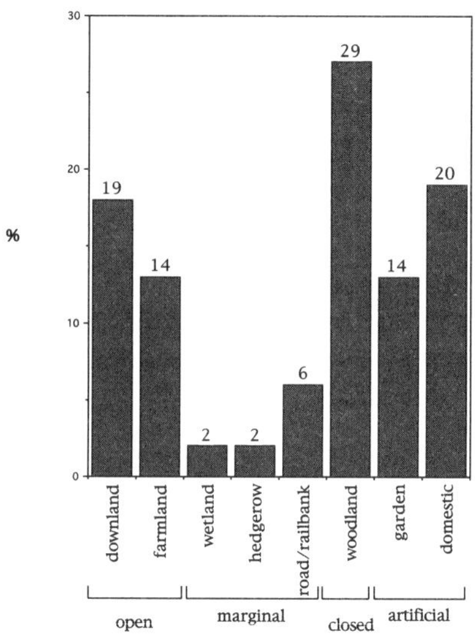

Fig. 24.3. Wood mouse: habitats

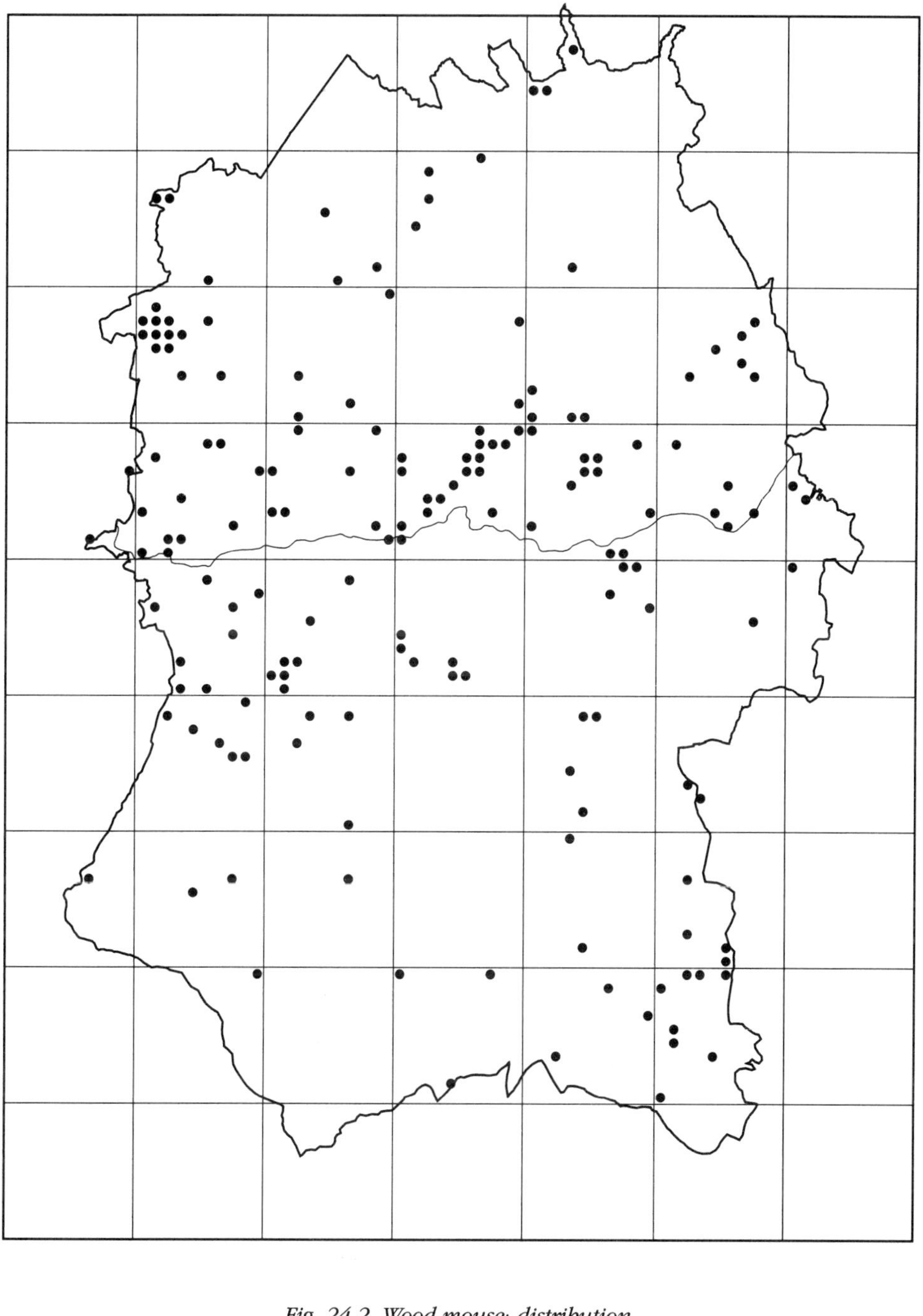

Fig. 24.2. Wood mouse: distribution

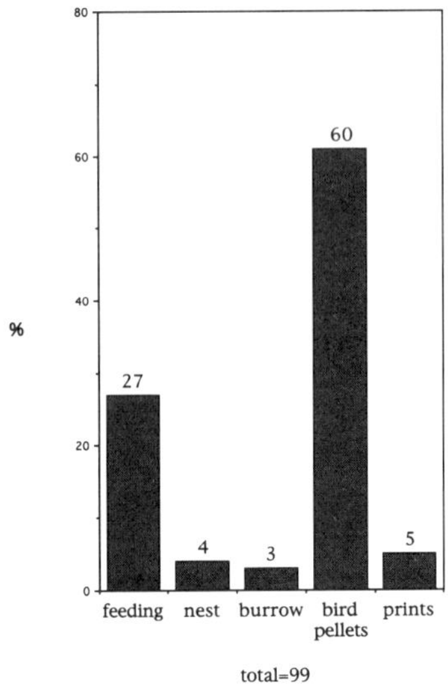

Fig. 24.4. Wood mouse: field signs

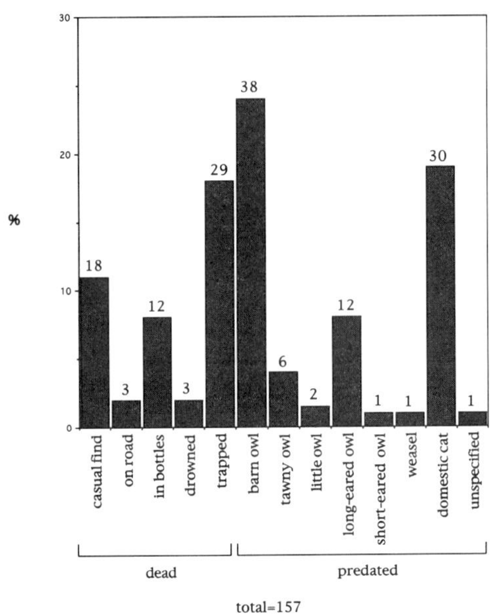

Fig. 24.5. Wood mouse: mortality

Order Rodentia — Family Murinae

Yellow-necked mouse — *Apodemus flavicollis*

Similar to wood mouse but slightly larger (fig. 21.1), with a yellowish chest patch, and slightly different foot and toe pads (fig. 21.2). Head and body 95-120 mm.

Handbook, 220-229; *Atlas*, 86-87; *Review*, 52-54; *Red Data Book*, 38-39.

Status

Native. Common where it occurs.

The total pre-breeding population in Britain is estimated at 750,000. The range of this species may have been more extensive in the past. The population is thought to be declining.

Nationally

Occurs mainly in southern England and Wales. Not recorded in the far north of England or in Scotland. Found mainly in woodland but may enter domestic buildings and outhouses.

Wiltshire

Data up to 1983 published in *WAM* 78, 94-104.

19 records, all of which have been used in this analysis. 63% of the records are of dead animals, 32% are of live animals and 5% are of field signs. Scattered distribution (fig. 25.1); gaps in the record where suitable habitat occurs are probably due to lack of recording and/or confusion with the wood mouse. All records are from woodland or buildings. The only record of a field sign is attributed to a nest. 12 records of mortality; of these, nine were due to traps and three due to predation by domestic cat.

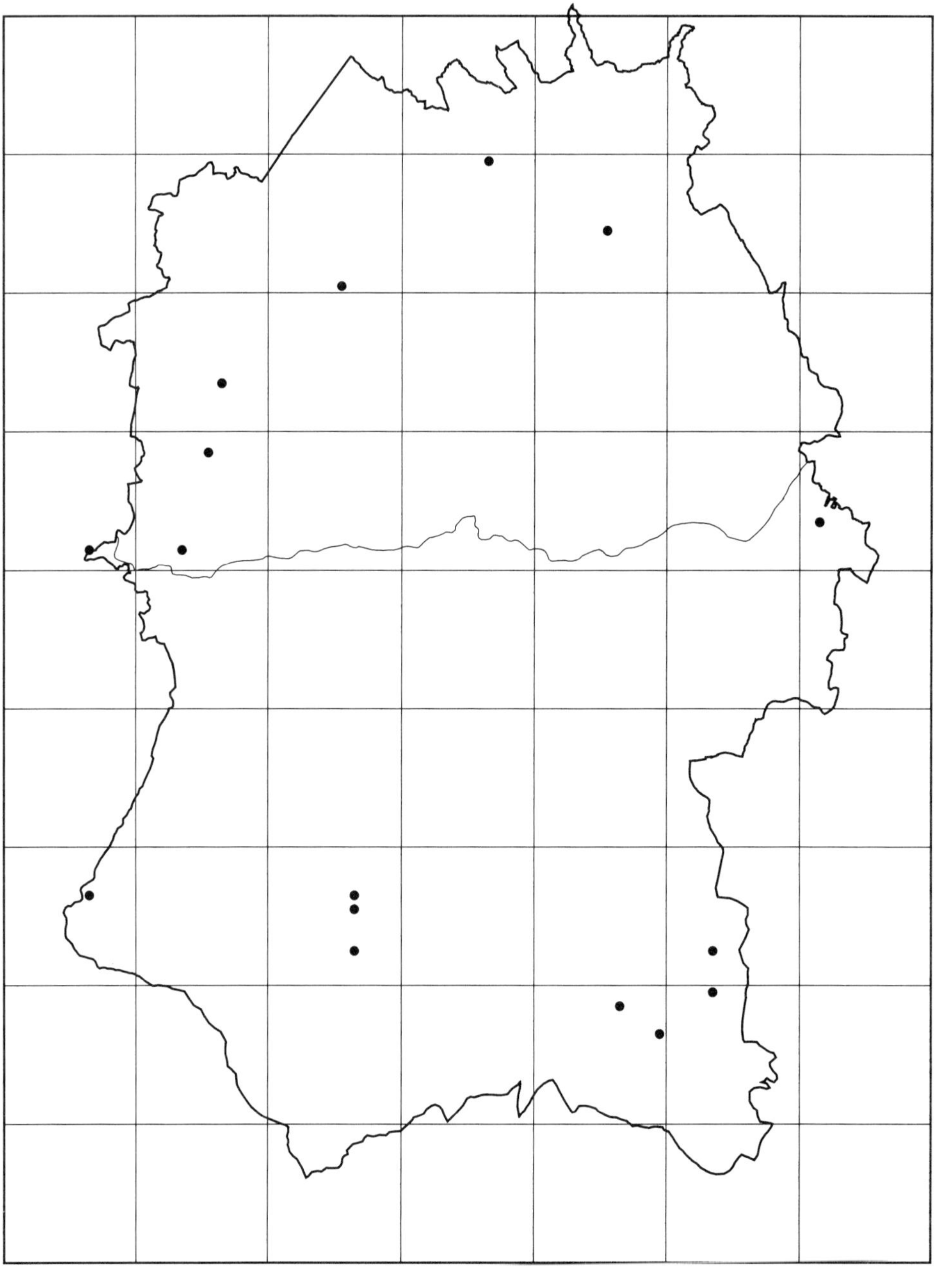

Fig 25.1. Yellow-necked mouse: distribution

Order Rodentia — Family Murinae
Harvest mouse — *Micromys minutus*

Very small mammal, superficially more like a vole than a mouse, with a blunt muzzle and small, hairy ears (fig. 21.1). The fur is russet-orange on the upper side of the body and pale, almost white on the underside. It is the only British rodent with a semi-prehensile tail. Head and body 50-70 mm. Foot and toe pads (fig. 21.2) and skull, jawbones and dentition (fig. 21.3) are diagnostic.

Handbook, 233-239; *Atlas,* 90-91; *Review,* 54-55.

Status

Native. Local. Common in places.

The total pre-breeding population in Britain is estimated at 1,425,000. Although generally designated a native, the harvest mouse may have been introduced in the Neolithic and its subsequent spread associated with the clearance of woodland and introduction of agriculture. It is now thought to be declining, mainly as a result of habitat loss.

Nationally

Widely distributed in southern England and the midlands. Present in parts of Wales and northern England. No recent records from Scotland. Occurs in a range of habitats, particularly open ones and those offering dense ground cover.

Wiltshire

First recorded by Montagu (1803). Hony (1917) claimed that Montagu was the first to record the species in England although the credit goes to Gilbert White who published the first account.

Data collected in Wiltshire as part of a national survey under the auspices of the Mammal Society published in Browne and Gillam (1974) and Dillon and Browne (1975). Data up to 1983 published in *WAM* 78, 94-104.

190 records, of which 182 have been used in this analysis. 81% of records are of field signs, the majority of which are of nests which are diagnostic and the means by which the presence of the species is most readily established (fig. 26.1). Widespread distribution (fig. 26.2); gaps in the record where suitable habitat occurs are probably due to lack of recording. Present in a wide range of habitats but predominantly downland and farmland (fig. 26.3). Wide range of mortality factors (fig. 26.4); the apparent predominance of predation by domestic cat and barn owl is probably a reflection of the data available from specific surveys ('what the cat brought in' (Ward 1981) and owl pellet workshops (see for example, Dillon, 1977)).

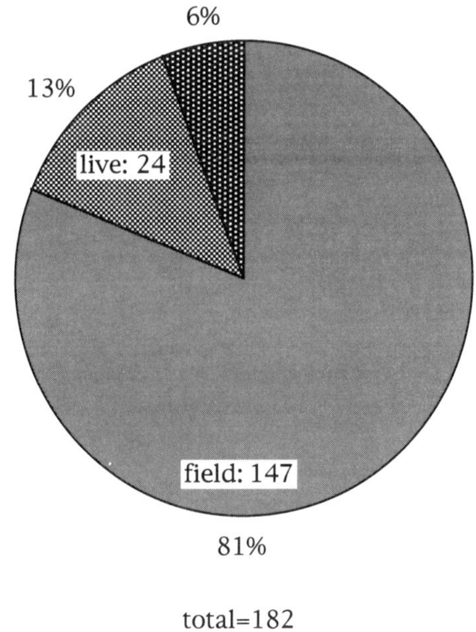

Fig. 26.1. Harvest mouse: number and nature of record

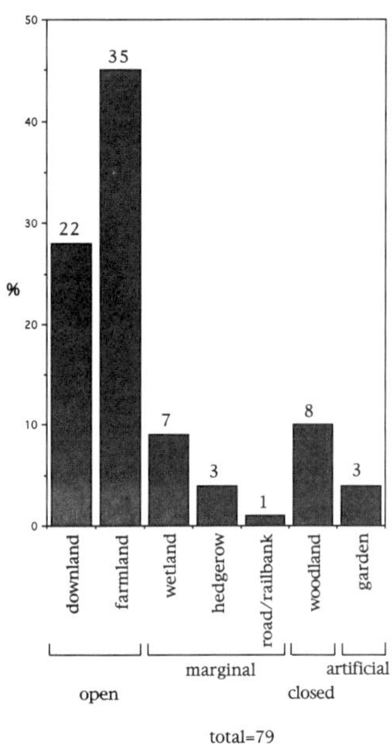

Fig. 26.3. Harvest mouse: habitats

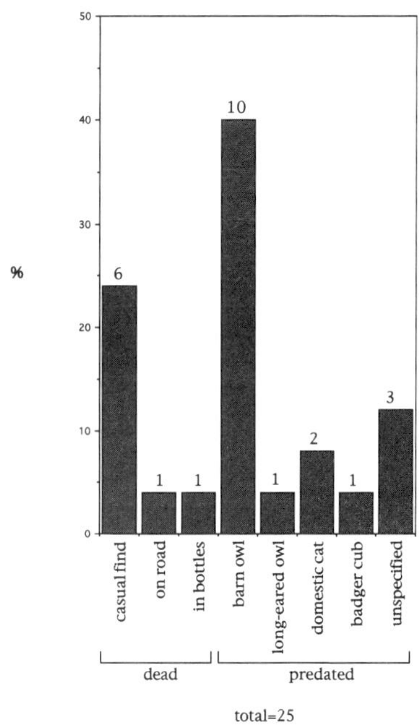

Fig. 26.4. Harvest mouse: mortality

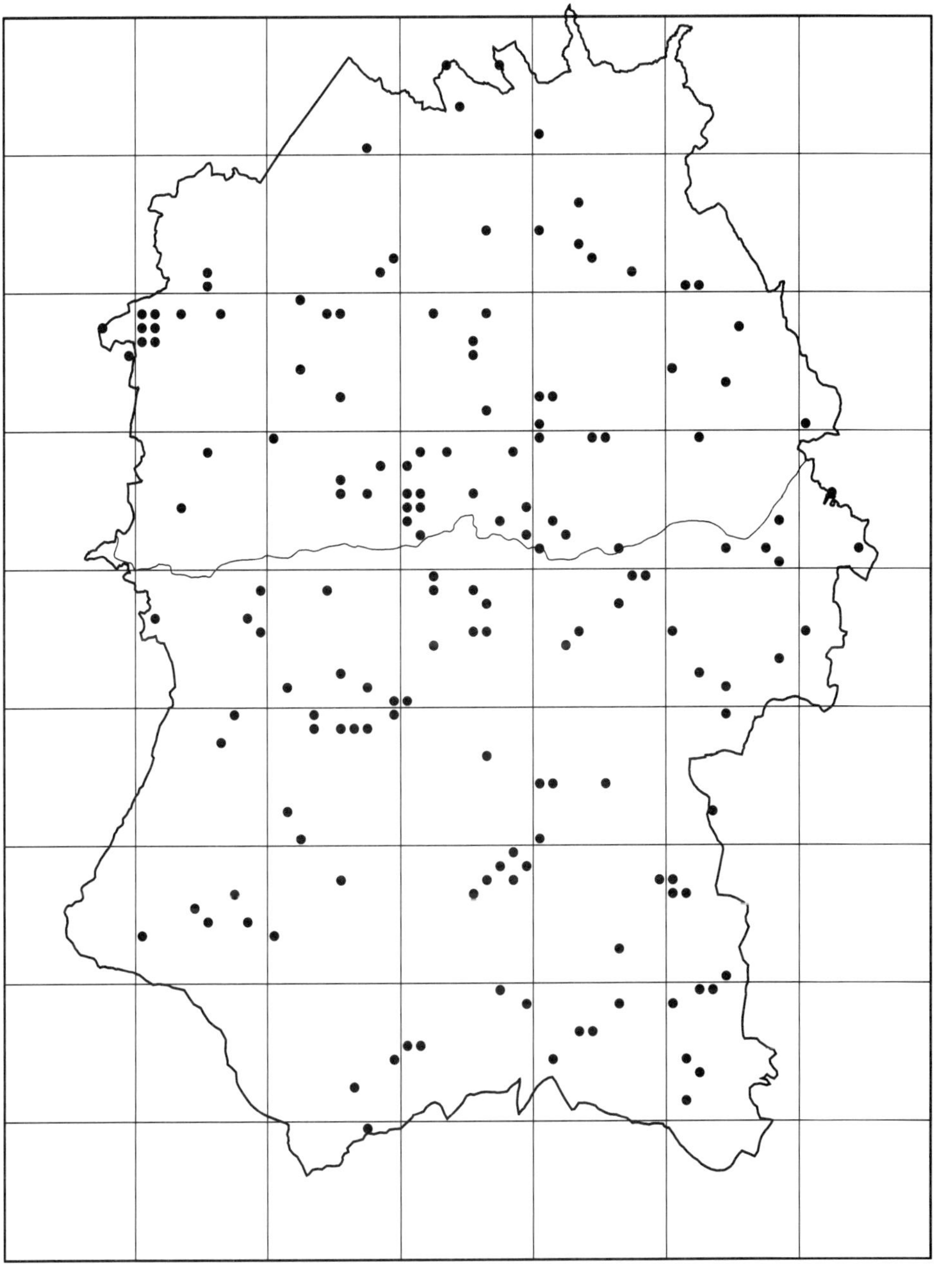

Fig. 26.2. Harvest mouse: distribution

Order Rodentia — Family Murinae

House mouse — *Mus domesticus*

Small mammal with a pointed muzzle and large ears (fig. 21.1). The fur is greyish-brown on the upperside, slightly paler on the underside. The presence of this species is often associated with a strong 'mousy' smell. Head and body 70-90 mm. Foot and toe pads (fig. 21.2) and skull, jawbones and dentition (fig. 21.3) are diagnostic.

Handbook, 239-247; *Atlas*, 92-93; *Review*, 56-58.

Status

Introduced. Common.

The total pre-breeding population in Britain is estimated at 5,192,000 although this may fluctuate widely. The population has declined in recent decades as a result of changing agricultural practice and pest control measures but it may now have stabilised.

Nationally

Widely distributed in England, Wales and Scotland. As it is commensal, it is most frequently, although not exclusively, encountered in association with human habitation. Probably under-recorded.

Wiltshire

Data up to 1983 published in *WAM* 78, 94-104.

93 records, of which 91 have been used in this analysis. 42% of records are of live animals (fig. 27.1). Widespread distribution (fig. 27.2); gaps in the record are probably due to lack of recording. Most records (17 out of 29) are associated with domestic buildings; there are small numbers of records for a range of other habitats. Remains in bird pellets account for most of the field signs (14 out of 17). Wide range of mortality factors (fig. 27.3); trapping and predation by domestic cat and barn owl predominate. The comparatively large numbers of records associated with the last two are probably a reflection of the data available from specific surveys ('what the cat brought in' (Ward 1981) and owl pellet workshops (see, for example, Dillon 1977)).

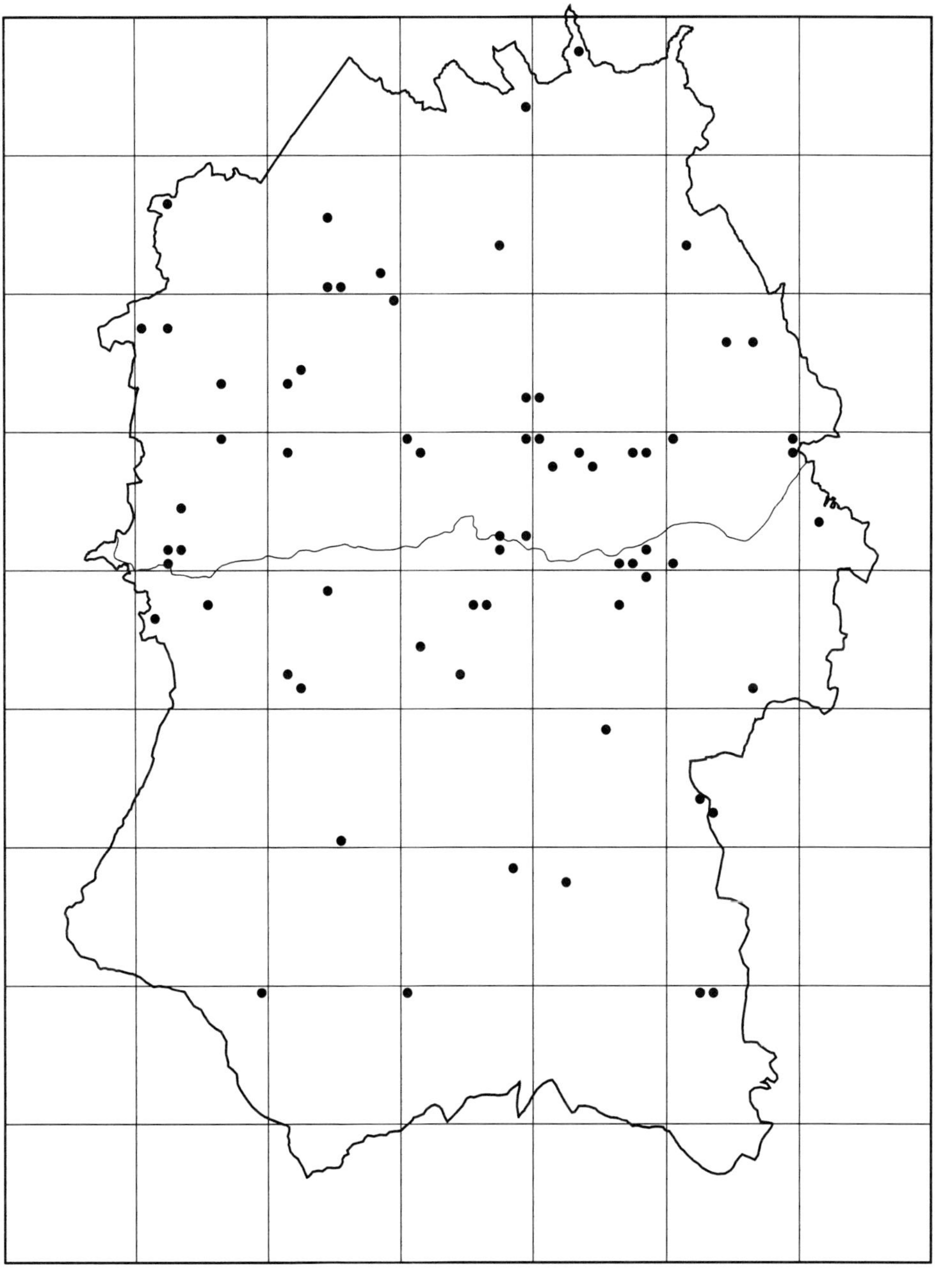

Fig. 27.2. House mouse: distribution

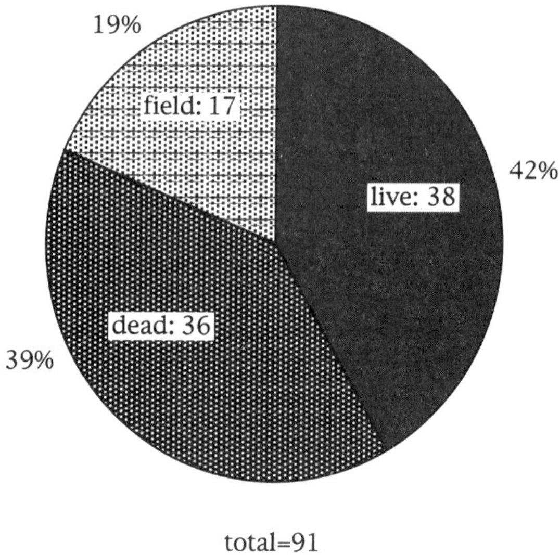

Fig. 27.1. House mouse: number and nature of record

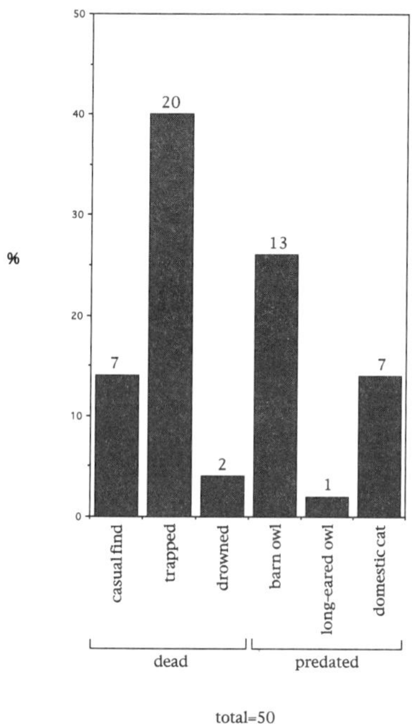

Fig. 27.3. House mouse: mortality

Order Rodentia — Family Murinae
Common rat — *Rattus norvegicus*

Also known as the brown rat. Small-medium sized mammal with a pointed muzzle, large ears and a long, scaly tail. Fur is greyish-brown on the upperside, paler on the underside. Features of the muzzle, ears, tail, foot and toe pads (fig. 23.2) distinguish it from the water vole. Head and body up to 280 mm. Skull, jawbones and dentition are diagnostic (fig. 28.1).

Handbook, 248-255; *Atlas,* 94-95; *Review,* 58-59; *Game Heritage,* 86-76.

Status

Introduced. Common.

The minimum pre-breeding population in Britain is estimated at 6,790,000 although this may fluctuate widely. The population has declined in number since its nineteenth century peak; continued decline may be attributed to changes in agricultural practice and pest control measures.

Nationally

Widely distributed in England, Wales and Scotland. As the rat is commensal, it is most frequently, although not exclusively, encountered in association with human habitation. Probably under-recorded.

Wiltshire

History of occurrence given in Dillon (1990). Special appeal for records made 1987. Data up to 1990 published in *WAM* 83, 190-200.

No local detail of earliest occurrence. Present in the eighteenth century, common in the nineteenth (Meyrick 1896) and early twentieth centuries (Hony 1917, Peirson 1923). Detailed observations over several months on a colony were made by Hawksley (1978).

405 records, of which 402 have been used in this analysis. 53% of records are of dead animals (fig. 28.2). Widespread distribution (fig. 28.3); gaps in the record are probably due to lack of recording. Present in a wide range of habitats (fig. 28.4). Wide range of field signs, remains in bird pellets and droppings predominating (fig. 28.5). Wide range of mortality factors and large number of records associated with these, particularly road casualties which account for 147 of the 229 records. (fig. 28.6).

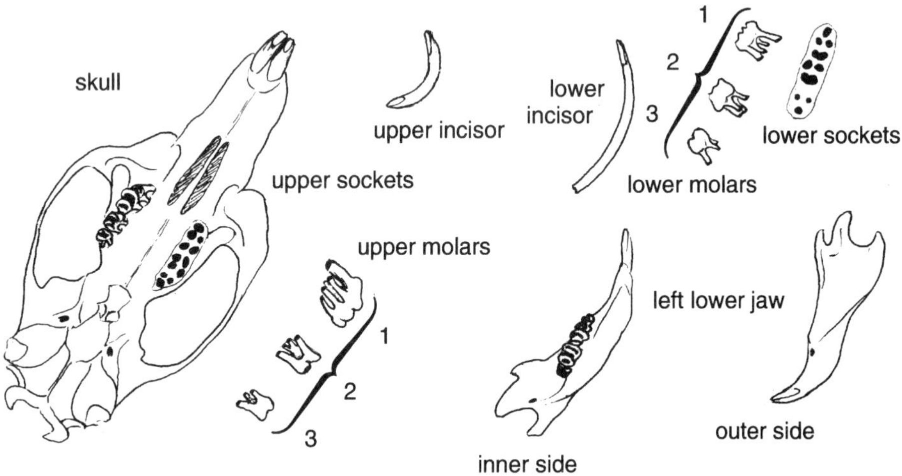

Fig. 28.1. Common rat: skull, jawbones and dentition, approximately x ⅔

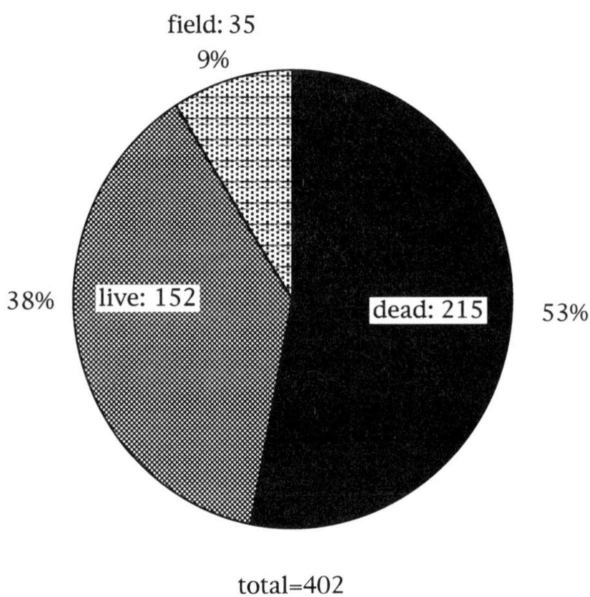

Fig. 28.2. Common rat: number and nature of record

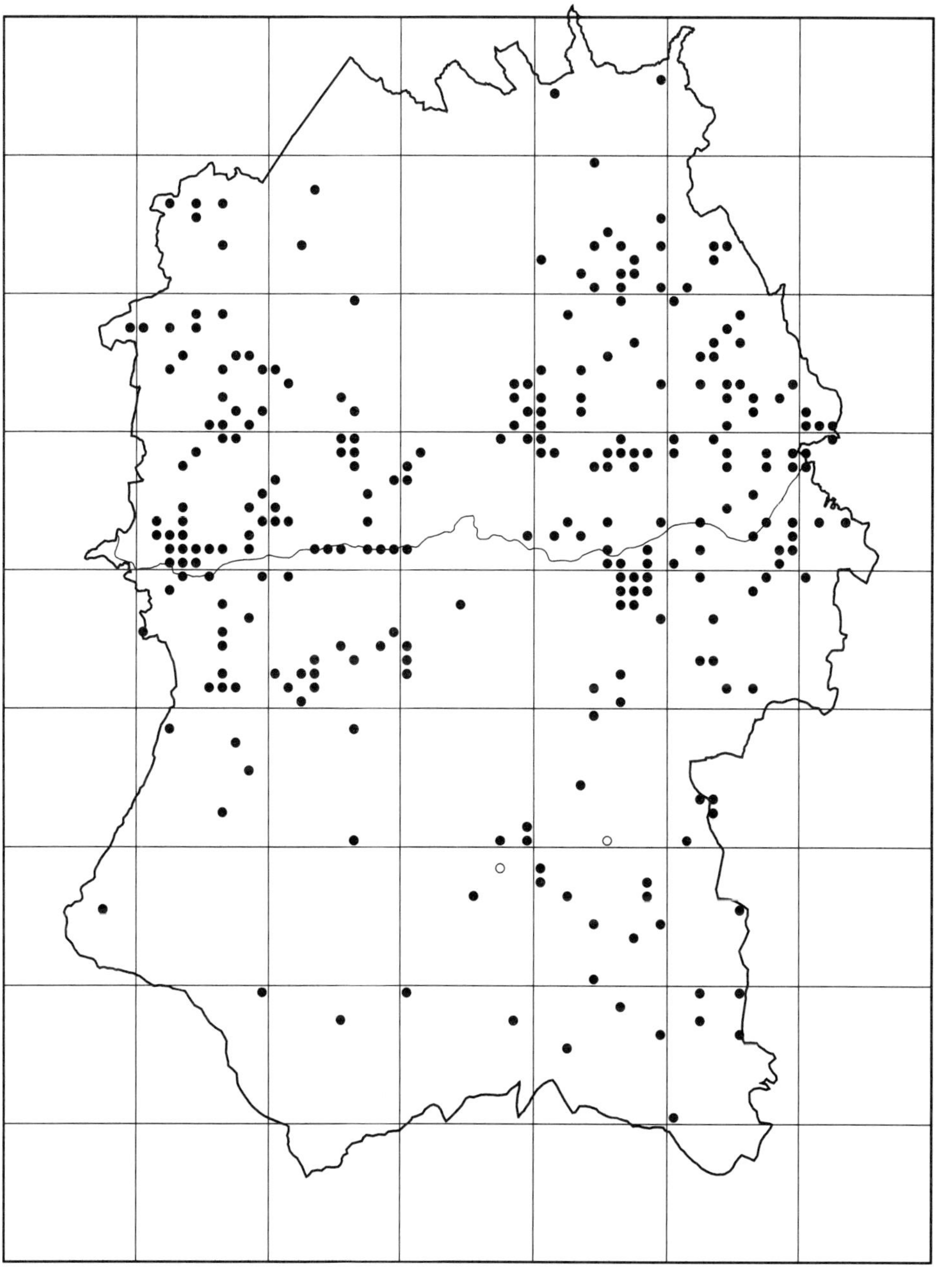

Fig. 28.3. Common rat: distribution

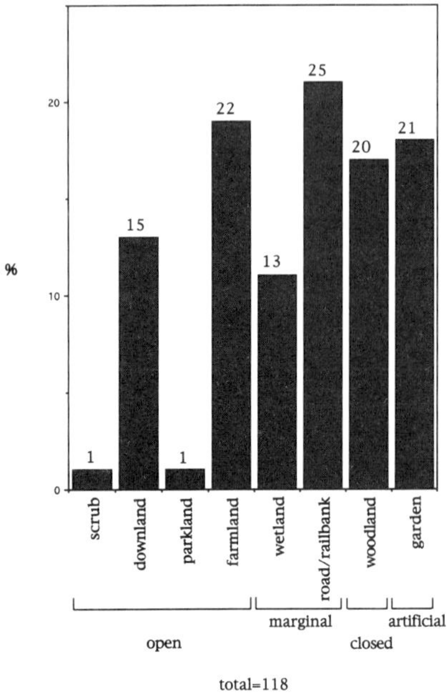

Fig. 28.4. Common rat: habitats

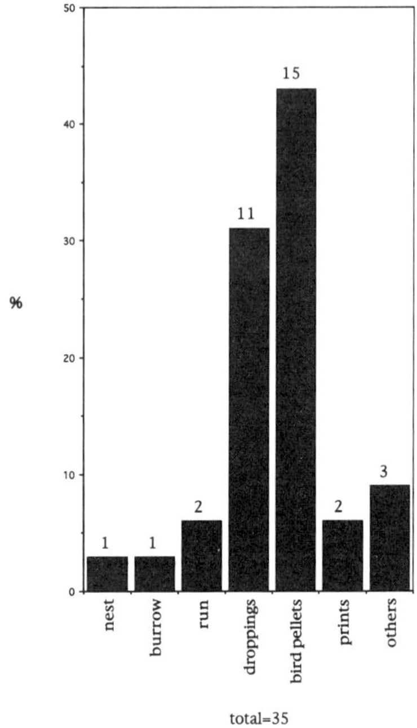

Fig. 28.5. Common rat: field signs

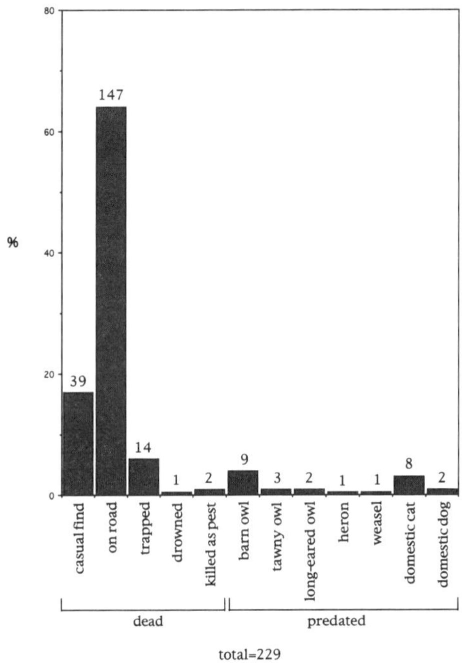

Fig. 28.6. Common rat: mortality

Order Rodentia — Family Murinae

Ship rat — *Rattus rattus*

Also known as the black rat. Small-medium sized mammal with large ears and eyes and a long, thin, tail. The colour of the fur varies: it may be grey-brown to black on the upperside and pale grey to creamy white on the underside. Head and body up to 240 mm.

Handbook, 1255-259; *Atlas,* 96-97; *Review,* 62-63; *Red Data Book,* 35-37.

Status

Introduced. Vulnerable and declining.

The total pre-breeding population in Britain is estimated at 1,300.

Nationally

Now reduced to a few scattered populations, mainly associated with sea ports, and on some off-shore Islands. As it is commensal, it is most frequently encountered in association with human habitation or activity.

Wiltshire

History of occurrence given in Dillon (1990). Present "in great abundance" in the eighteenth century (Montagu c.1785), thereafter it declined rapidly and was thought to be extinct by 1915 (Hony 1917). A population in farm buildings at West Kington was discovered in the 1970s and monitored until 1974 when the last individual was observed. This may have been a relict population or it may have been the result of translocation from an extant population elsewhere. There have been no records since that date and the species is now thought to be extinct.

Order Rodentia — Family Gliridae

Common dormouse — *Muscardinus avellanarius*

Small mouse-like mammal with golden or orange-brown fur and a bushy tail (fig. 29.1). Head and body 60-90 mm. Skull, jawbones and dentition diagnostic (fig. 29.2).

Handbook, 259-264; *Atlas*, 98-99; *Review*, 63-65; *Red Data Book*, 40-46.

Status

Native. Local and uncommon.

The total pre-breeding population in Britain is estimated at 500,000. The population has declined substantially since the nineteenth century; continued decline may be attributed to habitat loss, poor habitat management and the adverse influence of the weather.

Nationally

Widely distributed where habitat is suitable in southern England and parts of Wales, scattered distribution in the Midlands and northern England. No recent records from Scotland. The favoured habitat is deciduous woodland with plenty of understorey cover which includes hazel.

Wiltshire

Data up to 1984 published in *WAM* 79, 206-13. A survey was conducted by English Nature in 1995 as part of their species recovery programme (Bright 1996) but the data have not been used in this analysis.

66 records, of which 62 have been used in this analysis. 68% of records are of field signs (fig. 29.3). Widespread but localised distribution (fig. 29.4); gaps in the record where suitable habitat occurs are possibly due to lack of recording. Present in a small range of habitats but predominantly woodland (fig. 29.5). Comparatively large number of records are from domestic buildings. Nests, which are distinctive (fig. 29.6) account for 24 of the 42 records of field signs; feeding signs (fig. 29.7) account for a further 17 records. Very little information on mortality; of four records, two are of casual finds and one each are of an animal dead in a bottle and remains in a tawny owl pellet.

Fig. 29.1. Common dormouse: physical proportions, x ½

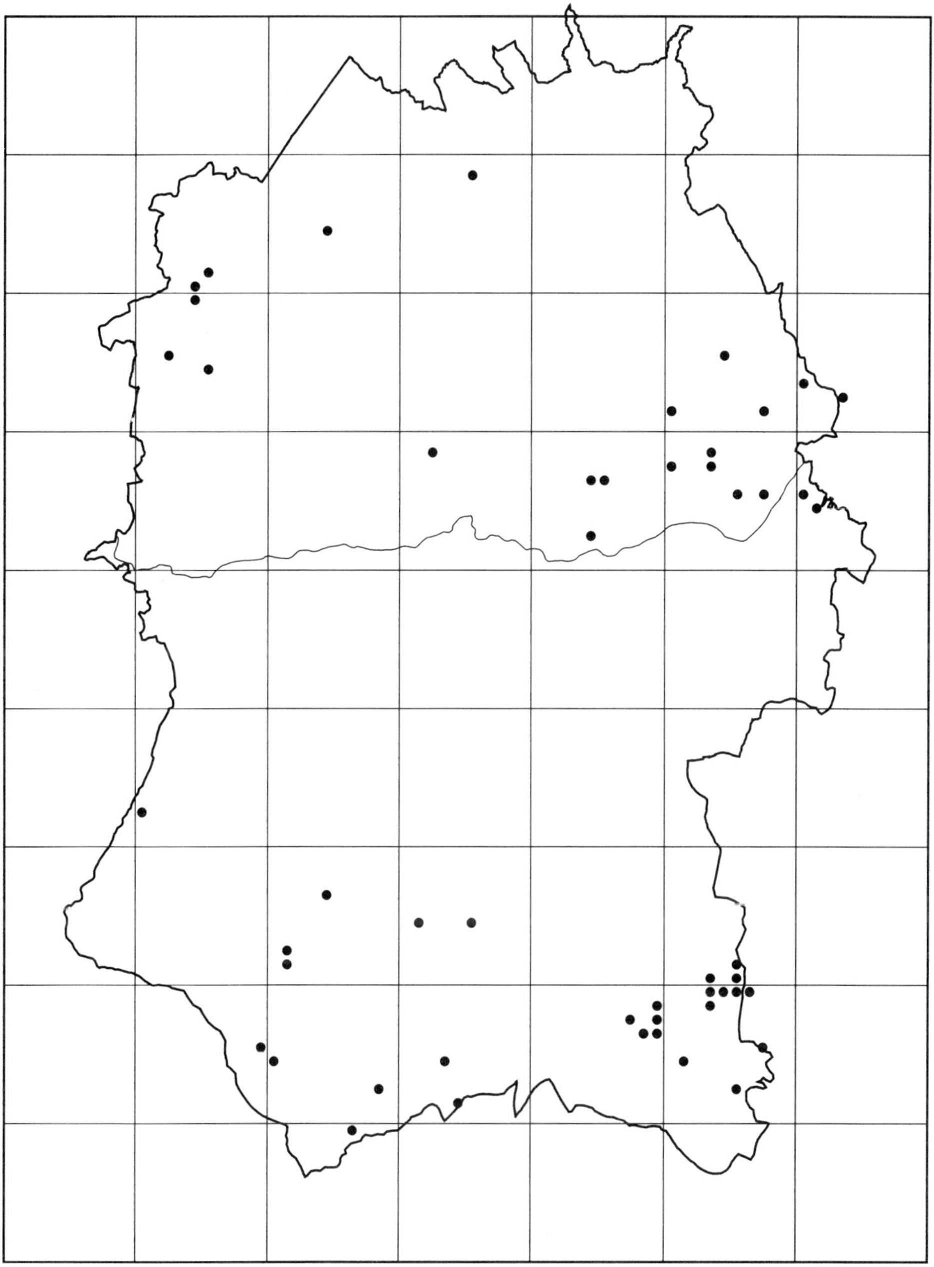

Fig. 29.4. Common dormouse: distribution

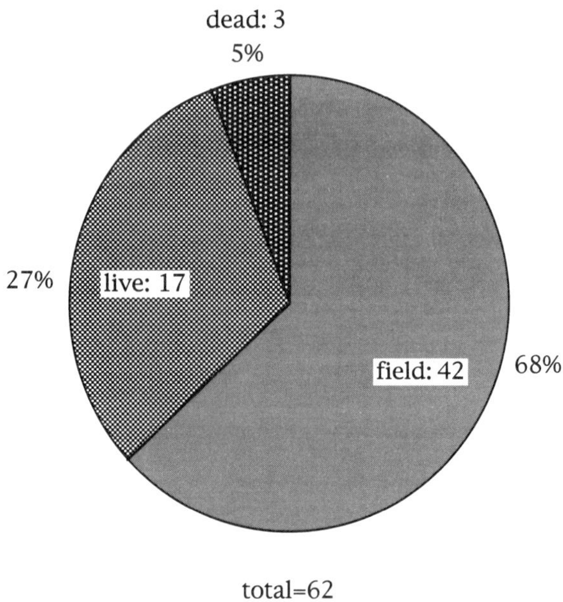

Fig. 29.3. Common dormouse: number and nature of record

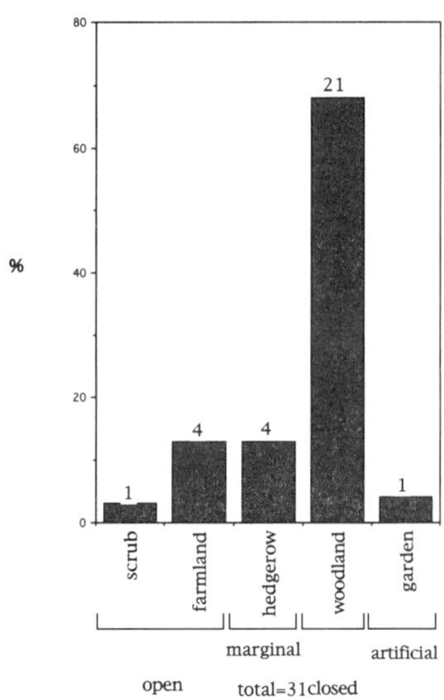

Fig. 29.5. Common dormouse: habitats

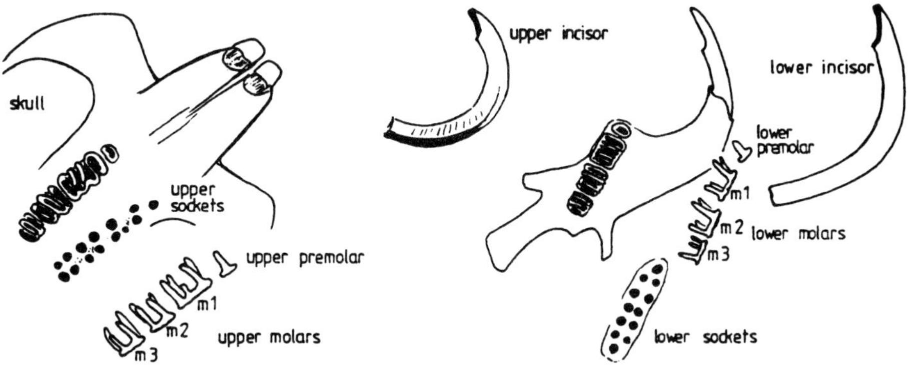

Fig. 29.2. Common dormouse: skull, jawbone and dentition, x 3

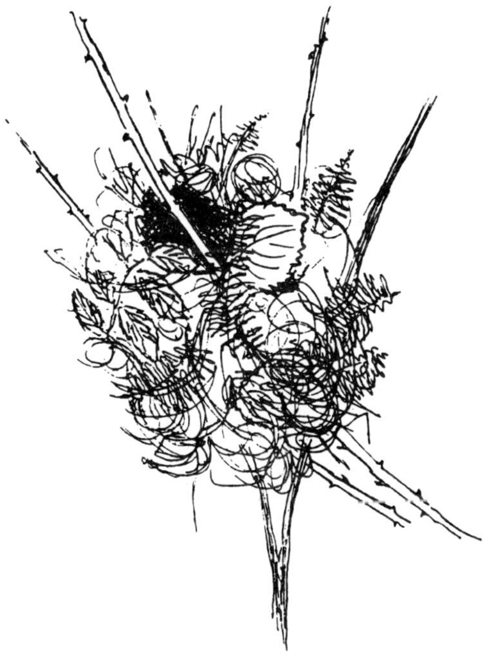

Fig. 29.6. Common dormouse: a typical nest, approximately x ½

Fig. 29.7: Hazel nuts opened by common dormouse, x 1

Order Rodentia — Family Capromyidae

Coypu — *Myocastor coypus*

Easily recognisable but poorly known medium sized mammal. Superficially rat-like but much larger with blunt muzzle, webbed hind feet and orange incisors. Head and body about 600 mm.

Handbook, 267-275; *Atlas*, 102-103; *Review*, 66-67; *Game Heritage*, 84-85.

Status

Introduced. Controlled and now probably extinct.

Nationally

Introduced in the 1930s for fur farming; animals escaped and several populations were established in the wild, notably in East Anglia. A systematic campaign of elimination began in the 1980s.

Wiltshire

A small population existed on the upper reaches of the River Kennet in the 1970s (Hawksley 1980) and there are records for 1981 and 1983. It is believed that none of these populations/individuals survive and it is presumed that the species is now extinct in the county.

Order Carnivora — Family Canidae

Wolf — *Canis lupus*

Distinctive, large member of the dog family with grey coat and long, bushy tail. Maximum height at shoulder 70-80 cm.

Status

Extinct.

Nationally

The historical evidence for the wolf in Britain is summarised by Harting (1880).

Wiltshire

Harting (1880) outlines his correspondence with A.C. Smith about the supposed association of wolves with the place name 'Wulfhall' in Savernake Forest and concludes that the link is spurious. The following records are from Hony (1917): "In 1321, William Mitchell held a messuage and land at Middleton Lilleborne [Milton Lilborne], Co. Wilts, of the King in capite, by the serjeanty of keeping his Wolf-dogs". In the *Inquisitions Post Mortem from the Reign of Edward III*, in the year 1328, it is recorded that "... the said land, pasture, wood, and rents [in Couelesfeld] are held by the sargeanty of taking wolves with the King's dogs at the King's expense when they come within the forest of Claryndone in Co. Wilts." In the same source, and referring to the year 1333, it is recorded that "...the tenements [in Coulsfeld Loueraz] are held on the King in chief by the service of hunting a wolf or wolves with the King's dogs through the whole county of Wilts, if any can be found in the county." These are the only known surviving references to wolves in Wiltshire. It is not known when the species became extinct in the county.

Order Carnivora — Family Canidae

Fox — *Vulpes vulpes*

Easily recognised, large, dog-like, carnivorous mammal, with yellow-brown to red-brown fur, erect, black-backed ears, a slender muzzle and a bushy-tail often with a white tip. Head and body 570-775 mm.

Handbook, 351-367; *Atlas,* 104-105; *Review,* 67-70; *Game Heritage,* 82-83.

Status

Native. Common.

The total pre-breeding population in Britain is estimated at 240,000. The fox has extended its range in the twentieth century and its numbers may still be increasing.

Nationally

Widely distributed in England, Wales and Scotland, under-recorded in some regions. A highly adaptable animal which utilises a wide range of habitats.

Wiltshire

The earliest records are from vermin accounts in 1752 (Smith 1879) and 1766 (Clay 1928).

No special survey of the fox has been published.

1018 records, of which 828 have been used in this analysis. 67% of records are of live animals (fig. 30.1). Widespread distribution (fig. 30.2); gaps in the record are probably due to lack of recording. Present in a wide range of habitats (fig. 30.3). A range of field signs has been recorded, particularly tracks, droppings and 'calls' (fig. 30.4). A range of mortality factors has been recorded (fig. 30.5): road casualties account for 92 of the 127 records; there are surprisingly few records of animals killed deliberately or through hunting although these are likely to be under-represented.

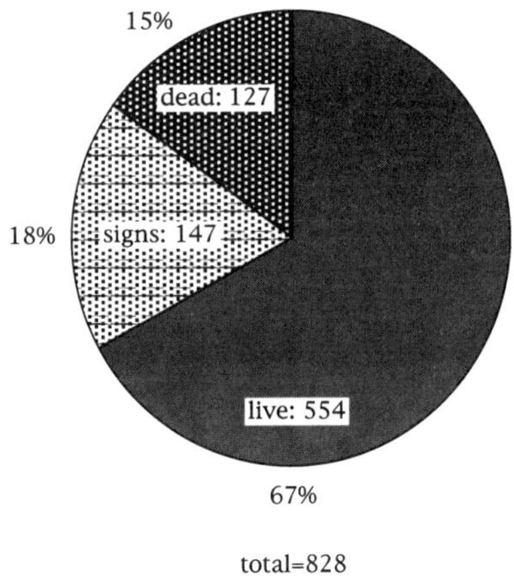

Fig. 30.1. Fox: number and nature of record

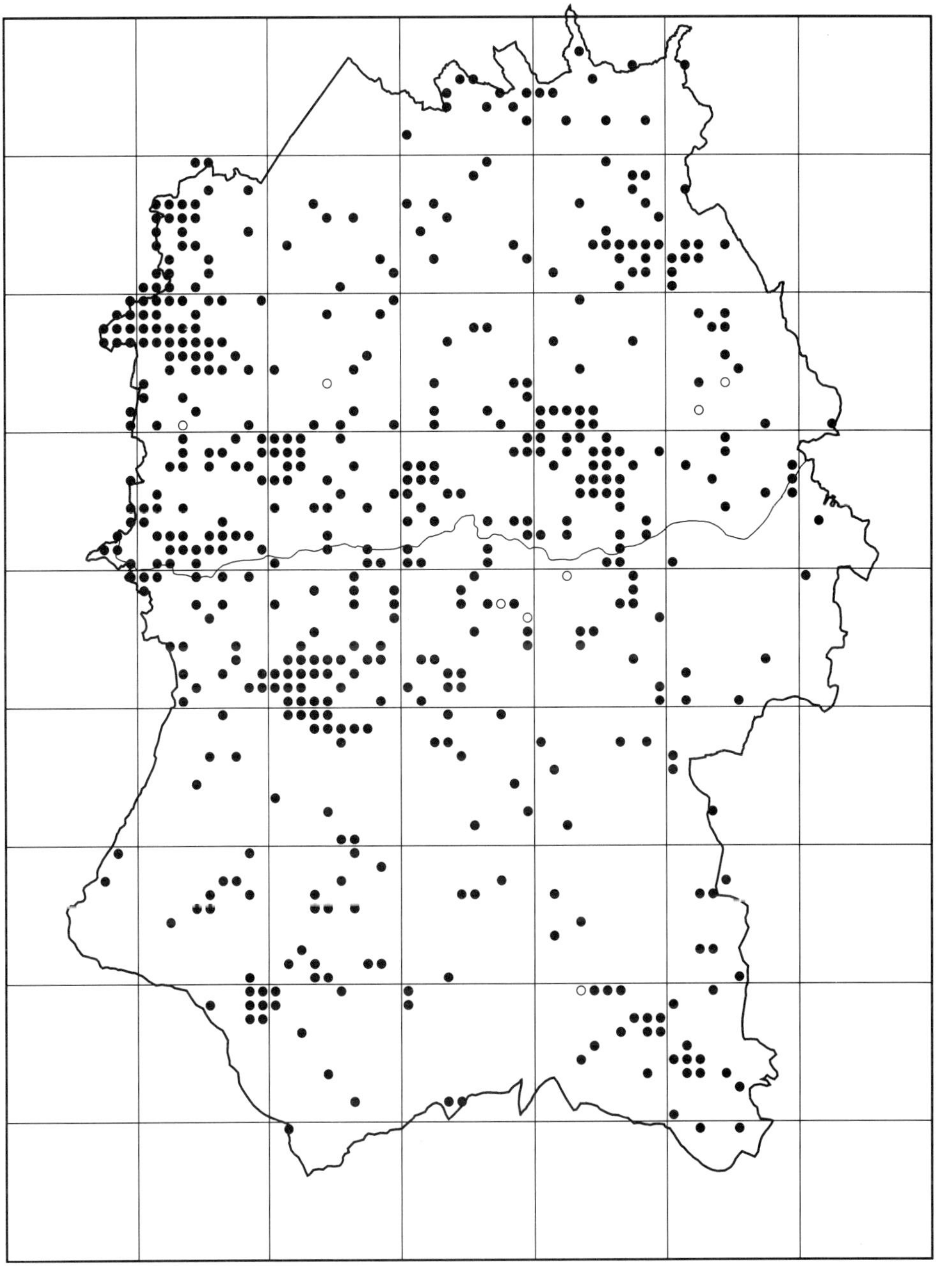

Fig. 30.2. Fox: distribution

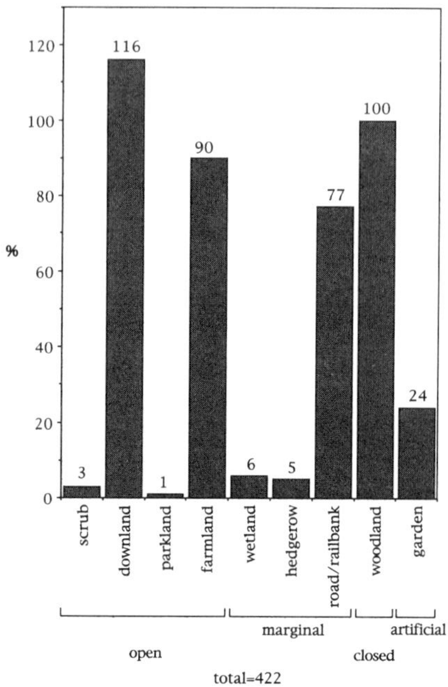

Fig. 30.3. Fox: habitats

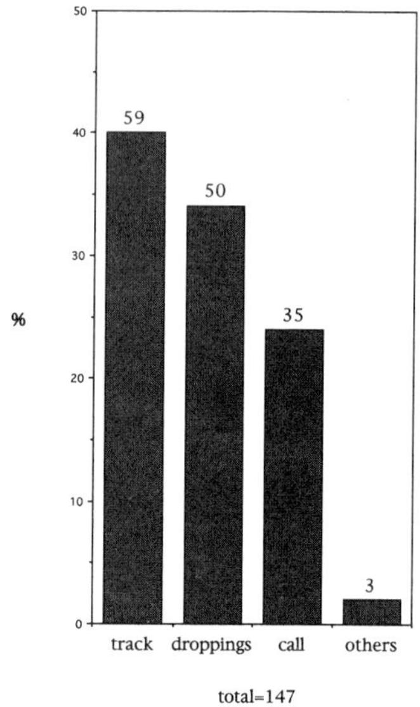

Fig. 30.4. Fox: field signs

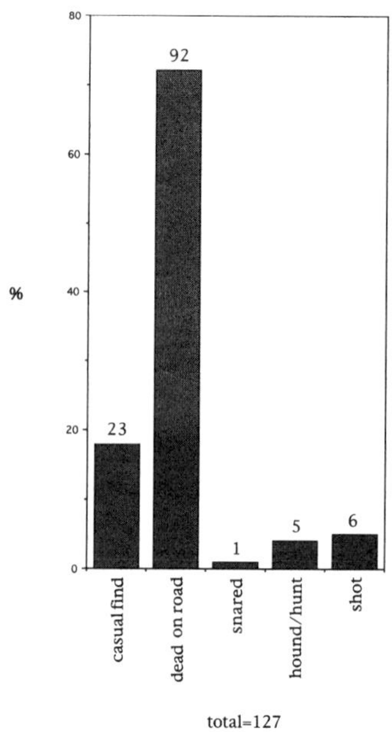

Fig. 30.5. Fox: mortality

Order Carnivora — Family Mustelidae

Pine marten — *Martes martes*

Medium-large mammal, with a long, slender, cat-like body, dark brown fur with a cream or yellow throat patch and long bushy tail. Head and body 410-540 mm.

Handbook, 368-376; *Review*, 70-72; *Atlas*, 106-107; *Red Data Book*, 76-77.

Status

Native. Uncommon.

The total pre-breeding population in Britain is estimated at 3,650. Until persecution in the nineteenth century reduced its range and numbers, the pine marten was widely distributed.

Nationally

The main centre of population is in the Scottish Highlands; there are scattered populations elsewhere in Scotland and some in England and Wales. The species may be extending its range where there is suitable habitat which in Britain is mainly coniferous and deciduous woodland.

Wiltshire

No longer present within the county.

Aubrey (c.1660) claimed that "upon these disafforestations the martens were utterly destroyed in North Wilts... In Cranborn Chase and Vernditch are some martens still remaining". The Cranborne Chase connection is corroborated by Chafin (1818): "packs of hounds have ever been kept in the neighbourhood of the Chase and hunted... martin-cats; but I believe the race of the latter are nearly extinct; their skins were too valuable for them to be suffered to exist". Pine martens were said by Millais (1904-6) to have been "exterminated as long ago as the end of the seventeenth century in Wilts." but Alfred Williams (1922) claimed that "the marten or martlet, survived in Stanton Woods and park down to the early part of the eighteenth century, though it is now quite extinct in the region of the Upper Thames". The 'Wiltshire Birds' papers at Devizes Museum contain a letter written to C. Rice in 1932 in which the correspondent claimed that one of his cases of stuffed animals included a pine marten which was shot in the 1820s or 1830s. There is no indication if this was a local specimen.

In 1953 a pine marten was reported near Grimstead. It was seen by several observers, many of whom reported it in detail (*SDNHS Report* 1953). No further sightings were reported in subsequent years. The animal may have escaped from captivity or have been deliberately released.

Order Carnivora — Family Mustelidae

Stoat — *Mustela erminea*

Medium sized mammal, with a long, slender body, red-brown fur on the upper side and a long tail with a black tip. Smaller than the polecat and larger than the weasel (fig. 31.1); foot and toe pads of the three species are diagnostic (fig. 31.2.). Head and body 240-310 mm.

Handbook, 376-387; *Atlas*, 108-109; *Review*, 72-73; *Game Heritage*, 74-75.

Status

Native. Common.

The total pre-breeding population in Britain is estimated at 462,000; it is thought to be declining.

Nationally

Widely distributed in England, Wales and Scotland; under-recorded in some regions. A highly adaptable animal which utilises a wide range of habitats but tends to avoid large open spaces.

Wiltshire

Aubrey (c.1660) mentions white stoats, ermines, at Vernditch Warren, otherwise early records are from vermin accounts (see, for example, Clay 1928).

No special survey of the stoat has been published.

299 records, of which 260 have been used in this analysis. 72% of records are of live animals (fig. 31.3). Widespread distribution (fig. 31.4); gaps in the record are probably due to lack of recording. Present in a wide range of habitats, particularly woodland and road/railbank (fig. 31.5). Of five records of field signs, three are of prints, one is of a dropping and one is unspecified. There are 67 records of mortality of which 35 are road casualties, 26 are of casual finds, four are of animals trapped and two are of animals taken by domestic cats; trapping is probably under-represented.

Weasel

Stoat

Fig. 31.1. Comparative sizes of weasel and stoat, approximately x ⅙

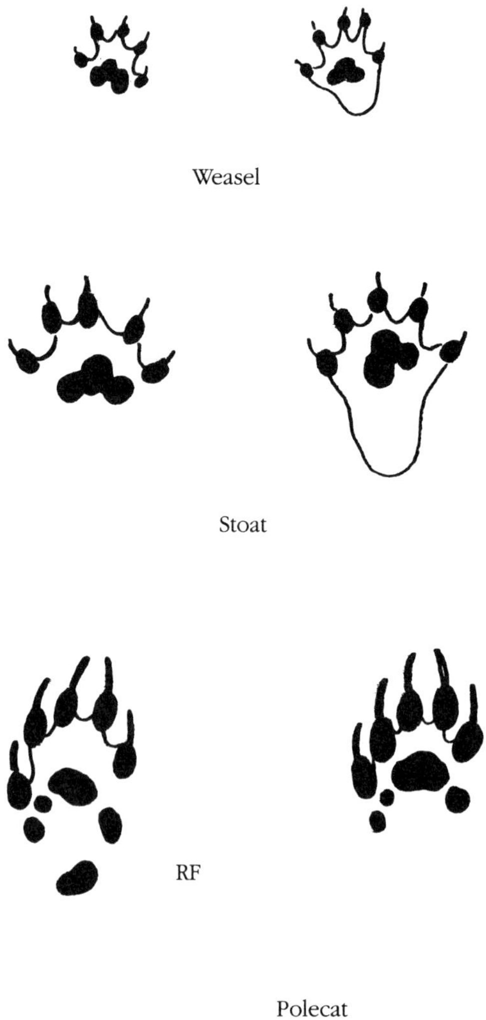

Fig. 31.2. Foot and toe pads of weasel, stoat and polecat, approximately x 1

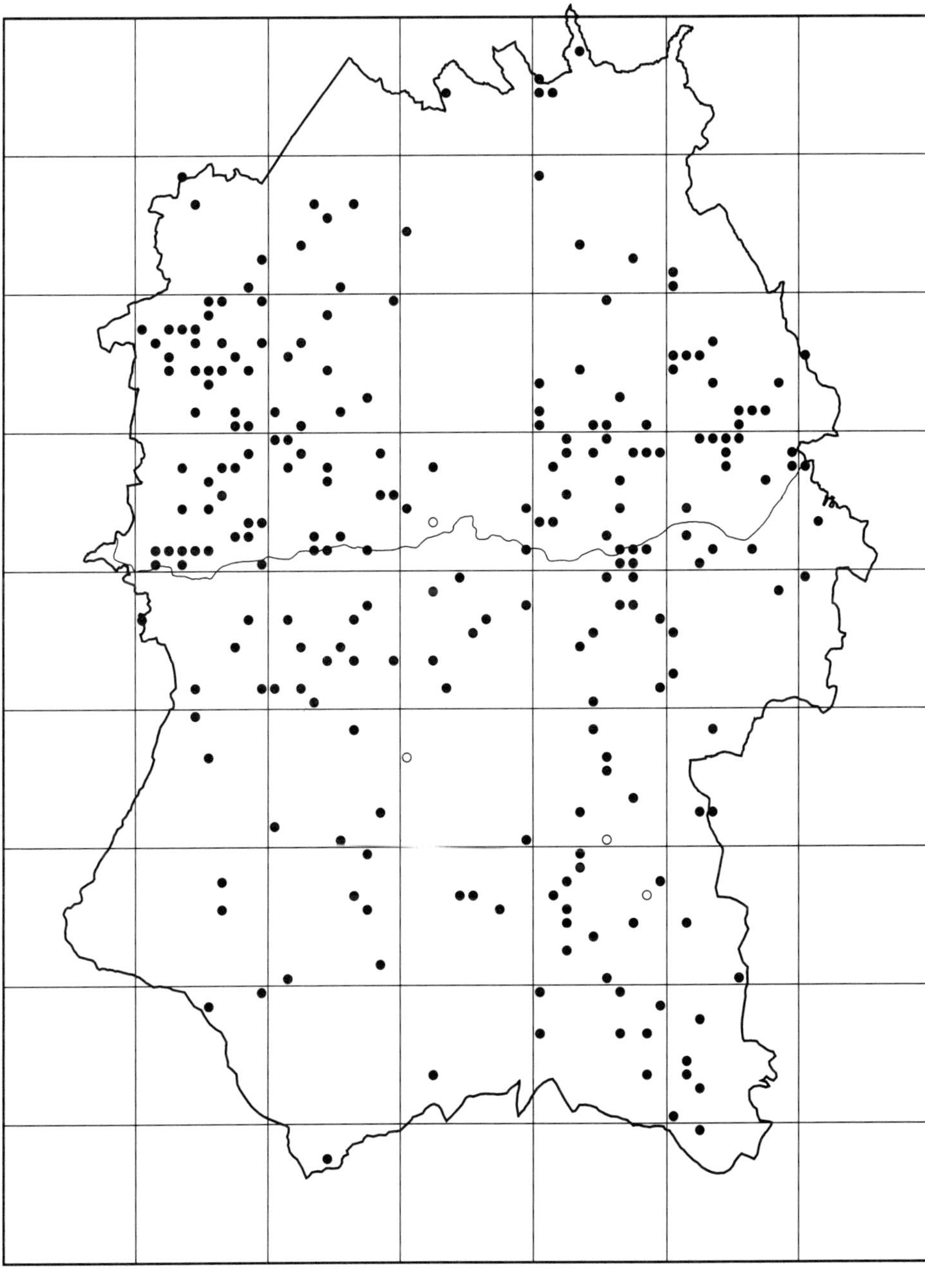

Fig. 31.4. Stoat: distribution

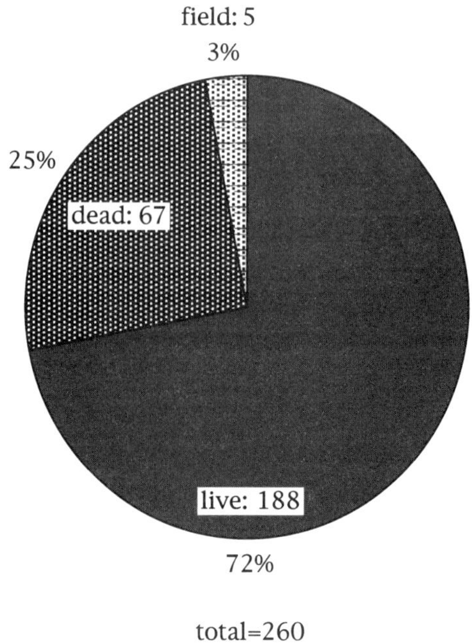

Fig. 31.3. Stoat: number and nature of record

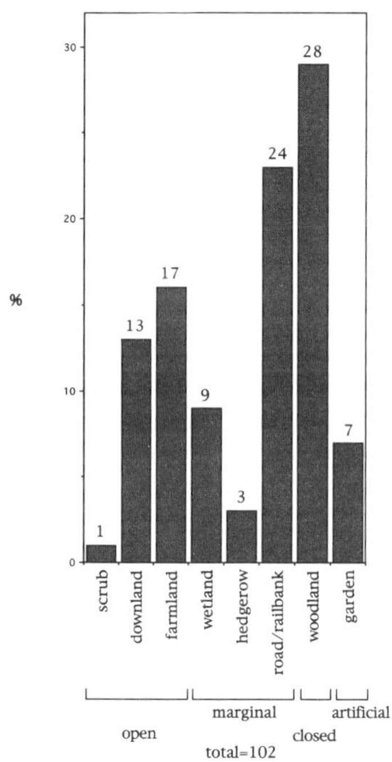

Fig. 31.5. Stoat: habitats

Order Carnivora — Family Mustelidae

Weasel — *Mustela nivalis*

Small carnivore, with a long, slender body, short legs and a short tail. The fur is usually ginger-brown on the upper side, white to cream on the lower side. Distinguished from the stoat by the lack of a black tip to the tail and smaller size (fig. 31.1). Foot and toe pads are diagnostic (fig. 31.2). Head and body 164-230 mm.

Handbook, 387-396; *Atlas*, 110-111; *Review*, 73-74; *Game Heritage*, 72-73.

Status

Native. Common.

The total pre-breeding population in Britain is estimated at 450,000; it is thought to be declining.

Nationally

Widely distributed in England, Wales and Scotland, under-recorded in some regions. A highly adaptable animal which utilises a wide range of habitats.

Wiltshire

No special survey of the weasel has been published.

357 records, of which 310 have been used in this analysis. 75% of records are of live animals (fig. 32.1). Widespread distribution (fig. 32.2); gaps in the record are probably due to lack of recording. Present in a wide range of habitats (fig. 32.3). Of two records of field signs, one is of a dropping and one is unspecified. There are 74 records of mortality, of which 37 are of animals killed as pests, 15 are of animals dead on roads, 13 are of casual finds, seven are of animals killed by domestic cats and one each are of an animal killed by a domestic dog and an animal drowned.

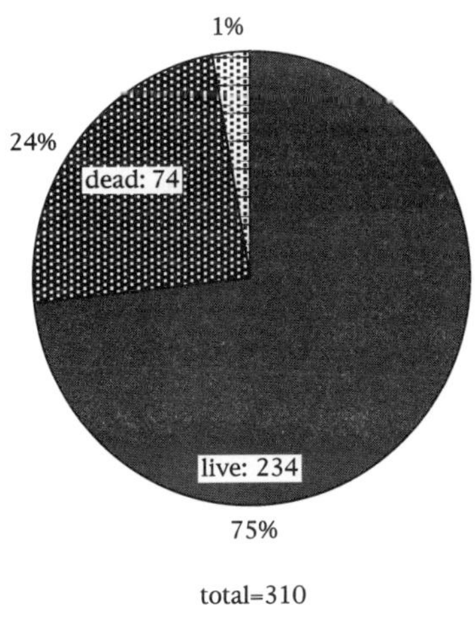

Fig. 32.1. Weasel: number and nature of record

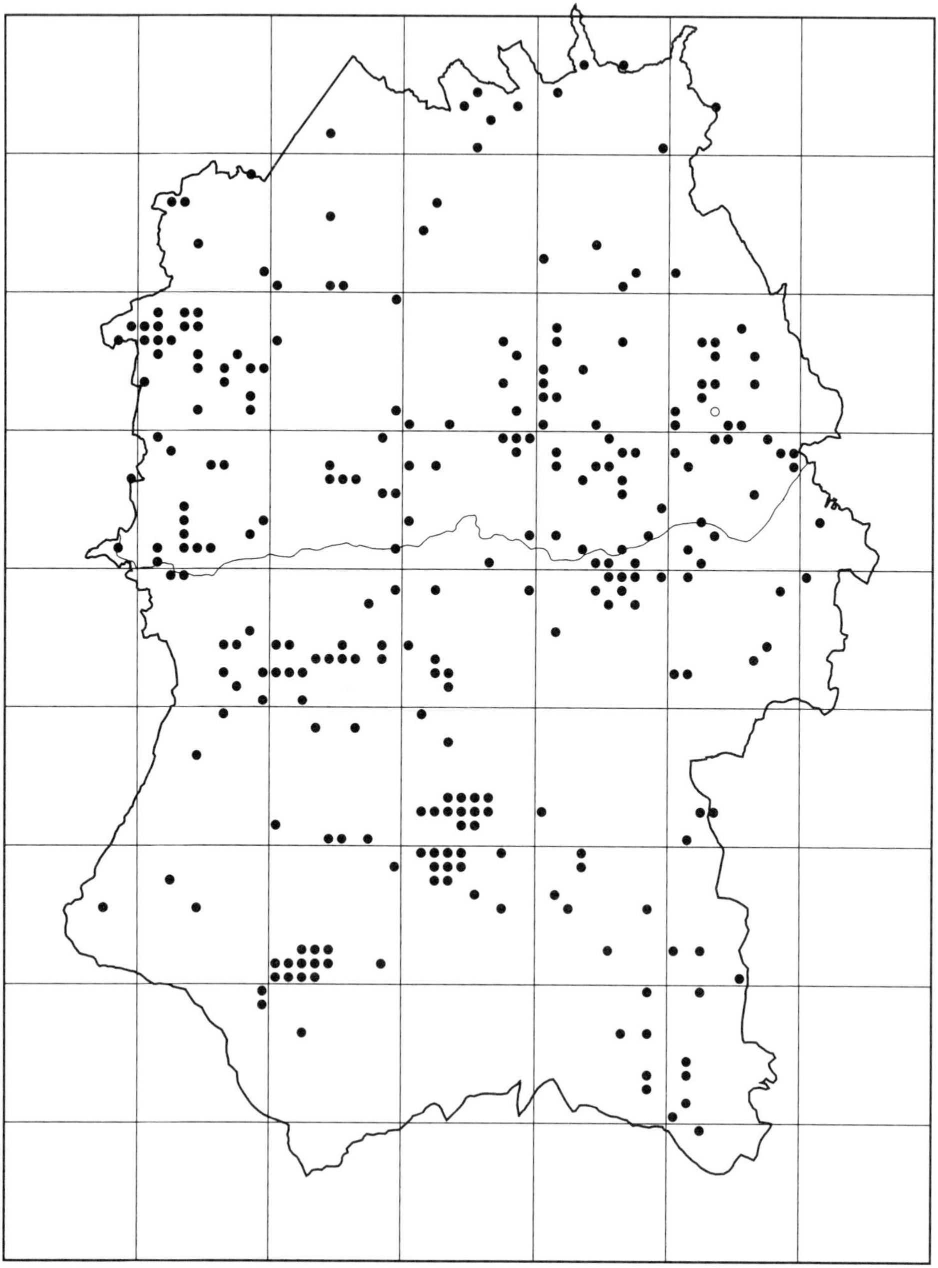

Fig. 32.2. Weasel: distribution

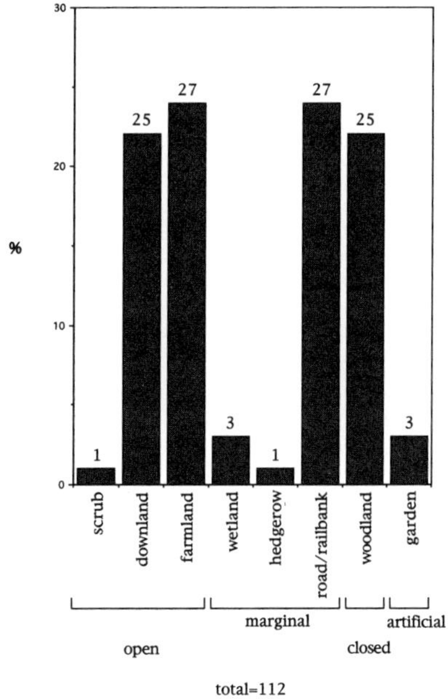

Fig. 32.3. Weasel: habitats

Order Carnivora — Family Mustelidae

Polecat — *Mustela putorius*

Medium sized mammal, with a long body, dark fur with distinctive light and dark facial markings. Very difficult to distinguish from dark coated feral ferrets. Head and body 330-450 mm.

Handbook, 396-405; *Atlas,* 112-113; *Review,* 74-76; *Red Data Book,* 80-83; *Game Heritage,* 76-77.

Status

Native. Restricted but may be common where it occurs.

The total pre-breeding population in Britain is estimated at 15,000 and increasing. Until persecution in the nineteenth century reduced its range and numbers, the polecat was widely distributed.

Nationally

Formerly widespread in Britain but now restricted to Wales and Welsh border counties where it is widely distributed in a range of habitats; extending its range.

Wiltshire

Present in Gloucestershire and the Upper Thames Valley (*Mammal News,* 104, 1996) and may now be present in north Wiltshire. However two road casualties in the south of the county, 1996, one near Stonehenge on the A344 and the other near Whiteparish, on the A36, have now been identified as true polecats (J. Birks pers. comm.). These records, together with others in the New Forest may result from an introduction.

Present in the nineteenth and twentieth centuries as the following records attest, although in some cases there may have been confusion with the feral ferret:

1792 – "Polecats were paid for by the Churchwardens at Yatesbury until about 1792" (*WAM* 18, 349).

1855 – "The Devizes Museum contains a pair killed in Castle Street Meadows, Salisbury, in 1855, and one presumably killed in Savernake Forest (no date)" (Hony 1917).

1855 – "... five polecats killed at Sutton Benger in 1855" (*WAM* 41, 429).

1870 – "One killed at Hilmarton about 1870" (Hony undated).

1872 – "... a correspondent remembers seeing a Polecat bolt from a rabbit hole in a small covert on Mauditt's Park Farm, near Malmesbury about the year 1872" (Hony 1917).

1886 – "A polecat has been trapped in the wood between Great Bedwyn and Chisbury ..." (*MCNHS Report* 34, 103).

1904 – "Four or five years ago I put one up by the river-side about half a mile from Melksham Church and in the same year I heard on reliable authority, of another between Shaw and Broughton Gifford. My late father has after told me that 60 years ago Polecats were fairly common in Melksham neighbourhood" (*WN&Q*, 4, 518).

1913 – "The polecat I saw during the last winter of 1913-14 near Marston Meysey, and I was close to it... in a wood lying between Marston Meysey and Fairford, there are, or were several polecats at least" (*WAM* 42, 80).

1921 – "Mr. Robert Newall of Fisherton House, has asked me to notify you of the occurrence of a polecat at Fisherton de la Mere on February 21st, 1921." There follows a detailed description and reasons why the author feels sure of his identification with notes to the effect that it was not a feral ferret. (*WAM* 41, 429).

1922 – "In a corner of a field, in which a large pile of loose thorn bushes has been stacked, I chanced upon a polecat with a small bird in its mouth" (Williams 1922).

Order Carnivora — Family Mustelidae

Feral ferret — *Mustela furo*

Medium sized mammal, with a long body. Fur colour varies: some individuals may be albino or have a pale coat; others have dark fur and facial markings which make them difficult to distinguish from the polecat. Head and body 330-450 mm.

Handbook, 405-406; *Atlas,* 114-115; *Review,* 76-77.

Status

Introduced.

The total pre-breeding population in Britain is estimated at 2,500.

Nationally

Widely distributed, scattered records in England, Wales and Scotland. It is not clear to what extent the records represent established populations.

Wiltshire

No special survey of the feral ferret has been published.

13 records, of which 12 have been used in this analysis. 75% of records are of live and 25% are of dead animals. Widespread and scattered distribution (fig. 33.1); records probably of escaped individuals with few, as yet, viable populations. Of the five records for which habitat information is available, two are from road/railbanks, one is from downland, one from wetland, and one from a garden. Of the three records for which mortality is recorded, two were of animals dead on roads and one was of an animal trapped.

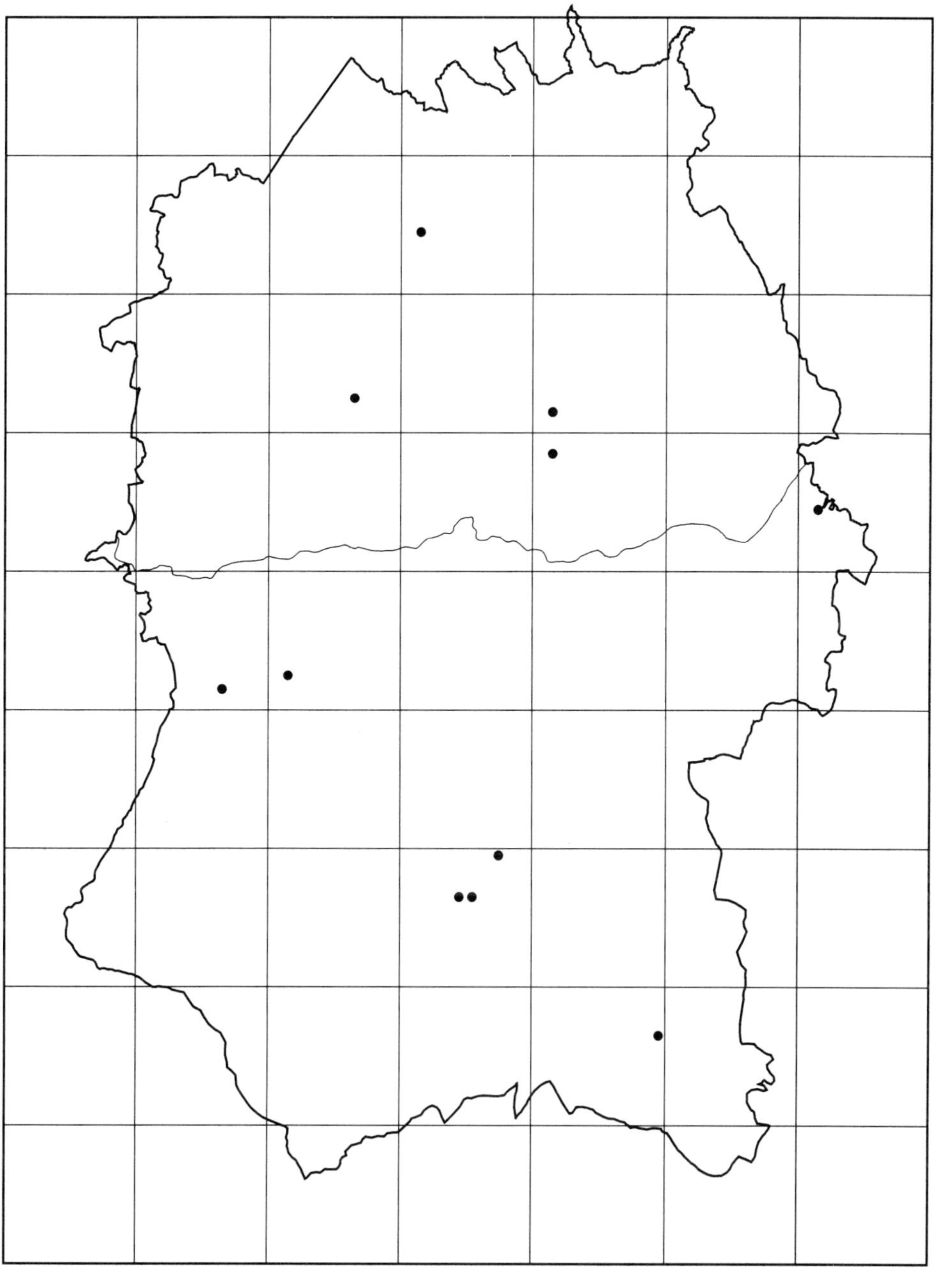

Fig. 33.1. Feral ferret: distribution

Order Carnivora — Family Mustelidae

Mink — *Mustela vison*

Medium sized mammal, with a long body. Fur colour varies but is usually dark brown with white patches under the chin. Head and body 330-450 mm.

Handbook, 406-415; *Atlas,* 116-117; *Review,* 77-80; *Game Heritage,* 78-79.

Status

Introduced. Increasing.

The total pre-breeding population in Britain is estimated at 110,000. The species is extending its range and increasing in number.

Nationally

Widely distributed in England, Wales and Scotland. Particularly associated with aquatic habitats.

Wiltshire

No special survey of the mink has been published.

The earliest records are from 1964-65. These, and all subsequent records until the 1970s, are from the Salisbury district. The *SDNHS Report* for 1967 records that 4,000 escaped from a farm near Stockbridge and there followed a rapid and widespread colonisation of the middle reaches of the Avon.

76 records, of which 61 have been used in this analysis. 51% of records are of live animals (fig. 34.1). Widespread but scattered distribution (fig. 34.2); most records are associated with the Salisbury Avon and Bristol Avon and their tributaries. Of 36 habitat records, 31 are of wetland, two are of farmland, two are of woodland and one is of a road/railbank. Of 24 records derived from field signs, 21 are of prints and three are of droppings. Of the six records for which mortality is recorded, two are of animals shot, two are of animals trapped, one is of an animal 'killed' and one is of an animal dead on a road.

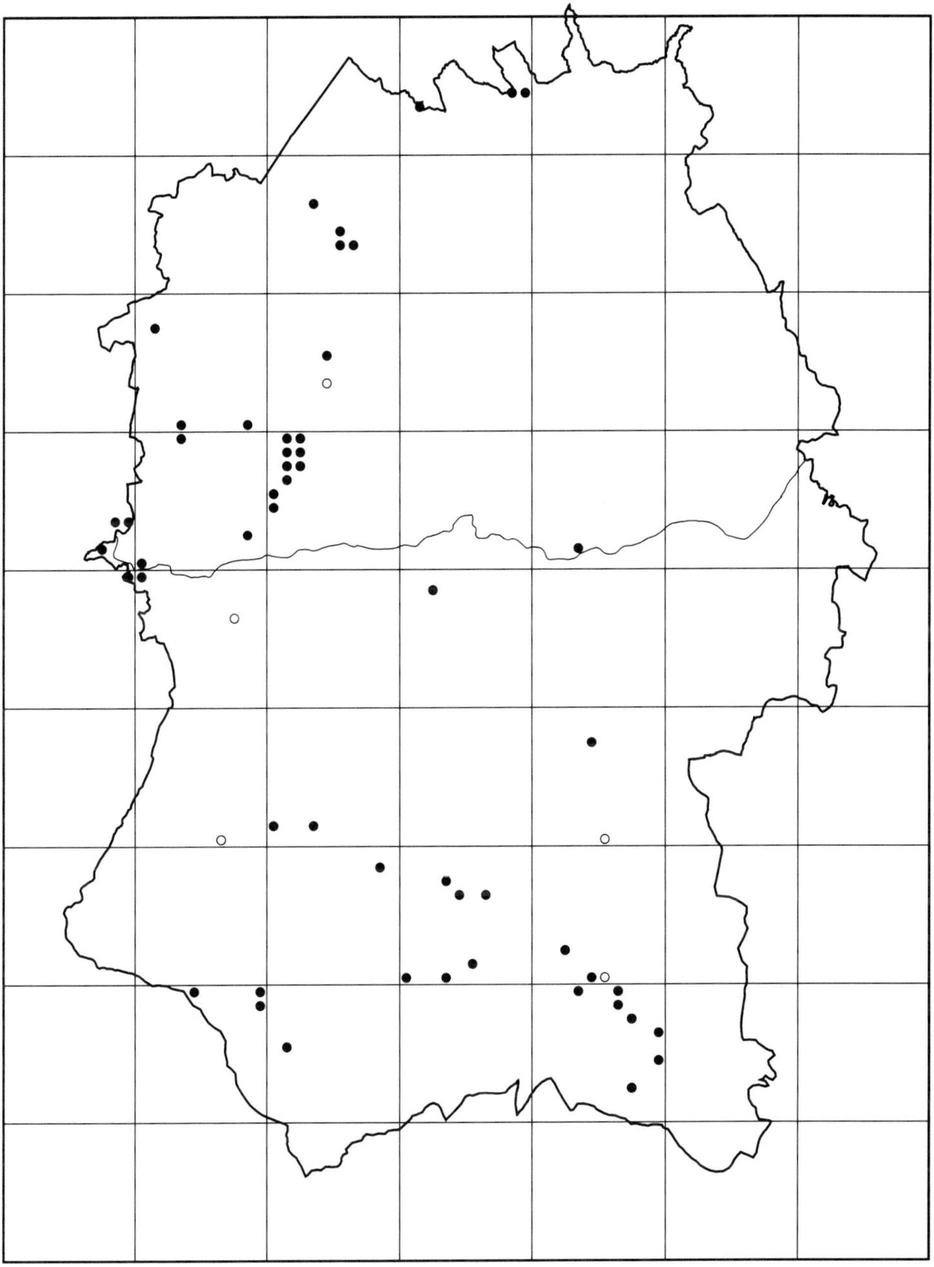

Fig. 34.2. Mink: distribution

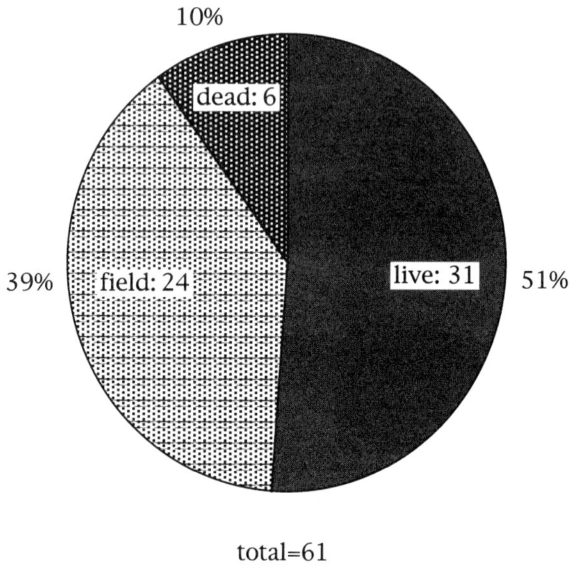

Fig. 34.1. Mink: number and nature of record

Order Carnivora — Family Mustelidae

Badger — *Meles meles*

Distinctive, large, powerfully-built mammal, with grey body fur and black and white striped head. Head and body 670-805 mm.

Handbook, 415-424; *Atlas*, 118-119; *Review*, 80-81; *Red Data Book*, 62-68.

Status

Native. Common.

The total pre-breeding population in Britain is estimated at 250,000. The status of the badger in Britain has been shown to be closely linked with changing patterns of agriculture (Reason, Harris and Cresswell 1993). The present population trends are unclear.

Nationally

Very widely distributed in England, Wales and Scotland; present in almost all 10 km squares in southern England, the Midlands and Wales. Found in a wide range of habitats.

Wiltshire

The status of the badger in Wiltshire is closely monitored by the Wiltshire Badger Group. A detailed survey was published in *WAM* in 1967 (Gillam 1967). For an analysis of casualties on Wiltshire roads see Scott-White (1991).

871 records, of which 536 have been used in this analysis. 57% of records are of field signs (fig. 35.1). Widespread distribution; gaps in the record where there is suitable habitat are probably due to lack of recording. Records come from all but five of the 10km squares of the County. These five are border squares with less than 50% of their area in Wiltshire. Present in a wide range of habitats with downland predominating (fig. 35.3). The majority of records of field signs are of burrows/setts (214 out of 306); 50 records are of tracks, 25 are of droppings, 5 are of animals 'calling', two are of prints and 10 are unspecified. There are 135 records of mortality of which 106 are of road casualties, 24 are of casual finds, 3 are of drowned animals; only two records are of animals killed deliberately although these are unlikely to be representative.

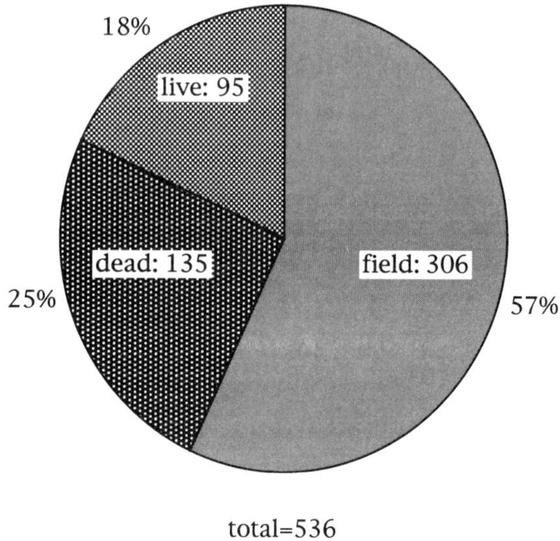

Fig. 35.1. Badger: number and nature of record

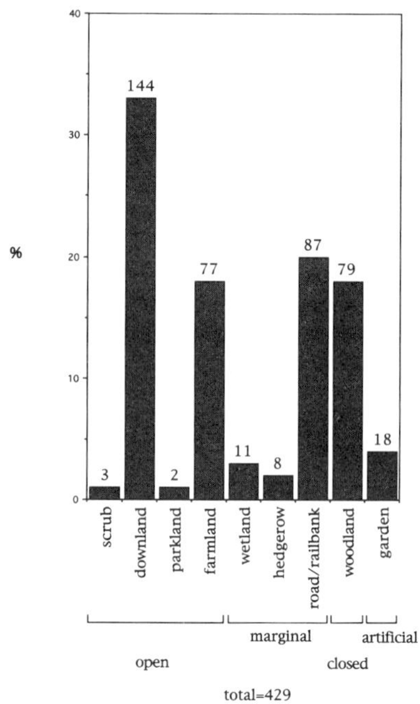

Fig. 35.3. Badger: habitats

Order Carnivora — Family Mustelidae

Otter — *Lutra lutra*

Medium-large, semi-aquatic, mammal, with a long, sleek body, small head, blunt muzzle and tapering tail. Head and body 630-730 mm.

Handbook, 423-431; *Atlas,* 120-121; *Review,* 81-85; *Red Data Book,* 69-95.

Status

Native. Vulnerable.

The total pre-breeding population in Britain is estimated at 7,350. The otter was widely distributed until the early nineteenth century. Thereafter, persecution, habitat loss and, in recent decades, pollution and road mortality have contributed to a substantial decline. The low point of the population was probably the late 1970s and there is now a slow recovery in some parts of the country assisted by some local restocking.

Nationally

Formerly widespread in Britain, the main centres of population are now restricted to Scotland, parts of northern England and Wales and south-west England. Absent from, or very uncommon in, southern, eastern and central England. Associated almost exclusively with wetland habitats.

Wiltshire

Present, but at very low population levels in a few localities only.

The earliest record for the county is from Aubrey (c.1660). There are records from the nineteenth century (Maton 1843, Scrope 1852, *MCHNS Report* 1886, and Meyrick 1896) and early twentieth century (Hony undated, Hony 1917). The species was said to be declining in the Upper Kennet region as early as the 1920s (Peirson 1923). There were regular reports in the Salisbury district, including the city, throughout the 1950s and early 1960s, and thereafter occasionally until 1976 *(SDNHS Reports).*

The status of the otter was unknown at the start of the present survey. Small populations were found on the Salisbury Avon and Bristol Avon river systems and these are thought to have been the nuclei of a small revival in numbers from about 1985. There have been some attempts at re-introduction but the details are confidential.

26 records, of which 24 have been used in this analysis. 54% of records are of live animals. There are records from the Salisbury Avon, Bristol Avon and Upper Kennet river systems. All 15 records of signs are of droppings (spraints). Of the two records of mortality, one was a casual find, the other was an animal dead on a road.

Order Carnivora — Family Felidae

Wild cat — *Felis silvestris*

The true wildcat looks like a large, solid, 'tabby'. It may be difficult to distinguish from feral cats with similar markings.

Handbook, 431-437; *Atlas,* 122-123; *Review,* 85-87; *Red Data Book,* 84-88.

Status

Native. Uncommon.

The total pre-breeding population in Britain is estimated at 3,500. Until persecution in the eighteenth and nineteenth centuries reduced its range and numbers, the wild cat was widely distributed. Hybridisation with domestic cats is thought to pose the greatest threat to its survival as a discrete species in Britain.

Nationally

The wildcat was once found throughout Britain but it is restricted to Scotland, mainly in the highland region.

Wiltshire

There are no records of the wildcat in Wiltshire but it almost certainly existed in the county until recent centuries. There are records from seventeenth and eighteenth century churchwardens' accounts in some Devon parishes (Dillon and Jones 1986).

There are feral cats in many parts of Wiltshire but as it is almost impossible to make a distinction between feral and domestic populations no systematic recording of the former has been undertaken.

Order Artiodactyla — Family Suidae

Wild boar — *Sus scrofa*

Large, distinctive member of the pig family. Maximum shoulder height 90 cm.

Status

Extinct since the seventeenth century.

Wiltshire

Aubrey (c.1660) gives an account of a sculpture over a Norman doorway at Little Langford Church depicting the hunting of wild boar. This is almost certainly a representation of an indigenous wild boar. Extracts from the account books of Edward Seymour of Wulfhall indicate that wild boar were killed in Savernake Forest around 1540 (Jackson 1875). At the time of Elizabeth I (1558-1603) there were wild boars in Grovely (Straton 1902); Sir Richard Grobham, who died in 1629 aged 78, "slew a wild boar in Grovely Woods, which was the terror of all the neighbourhood" (Wordsworth 1907). But Aubrey notes also that Charles I (1625-1649) had wild boar reintroduced to the New Forest from France because the indigenous stock had become extinct.

Order Artiodactyla — Family Cervidae

Red deer — *Cervus elaphus*

Large deer with a reddish-brown coat which, as an adult, does not have spots. The lack of black marking on the rump and tail is another distinguishing feature. Maximum shoulder height 120 cm.

Handbook, 492-504; *Atlas,* 128-129; *Review,* 95-97.

Status

Native. Common where it occurs.

The total pre-breeding population in Britain is estimated at 360,000 and is thought to be relatively stable.

Nationally

Widely distributed but the main centres of population are in moorland areas of Scotland and south-west England. There are scattered populations elsewhere often in woodland.

Wiltshire

No special survey of the red deer has been published.

Opinion is divided about the suitability of the large tracts of woodland in Wiltshire for red deer and the status of the herds which occur. Chafin (1818) argued that Cranborne Chase is not suitable for red deer and indeed none has been recorded there. On the other hand, it was claimed by Meyrick (1896) and Peirson (1923) that the small herd in Savernake Forest is descended from native wild animals. There are two records of lone animals associating with herds of cattle – one on Porton Ranges in the 1960s (*SDNHS Report* 1966) and the other on Fyfield Down in the 1970s.

12 records of which nine have been used in this analysis. Apart from two single records in the west of the county, the main clusters of records are in the extreme south-east bordering the New Forest and the area around Savernake Forest in the east (fig. 37.1). Of the nine records, seven are of live animals and two are of field signs, prints in both cases. Of seven habitats recorded, three are of downland, two of parkland and one each of woodland and roadside.

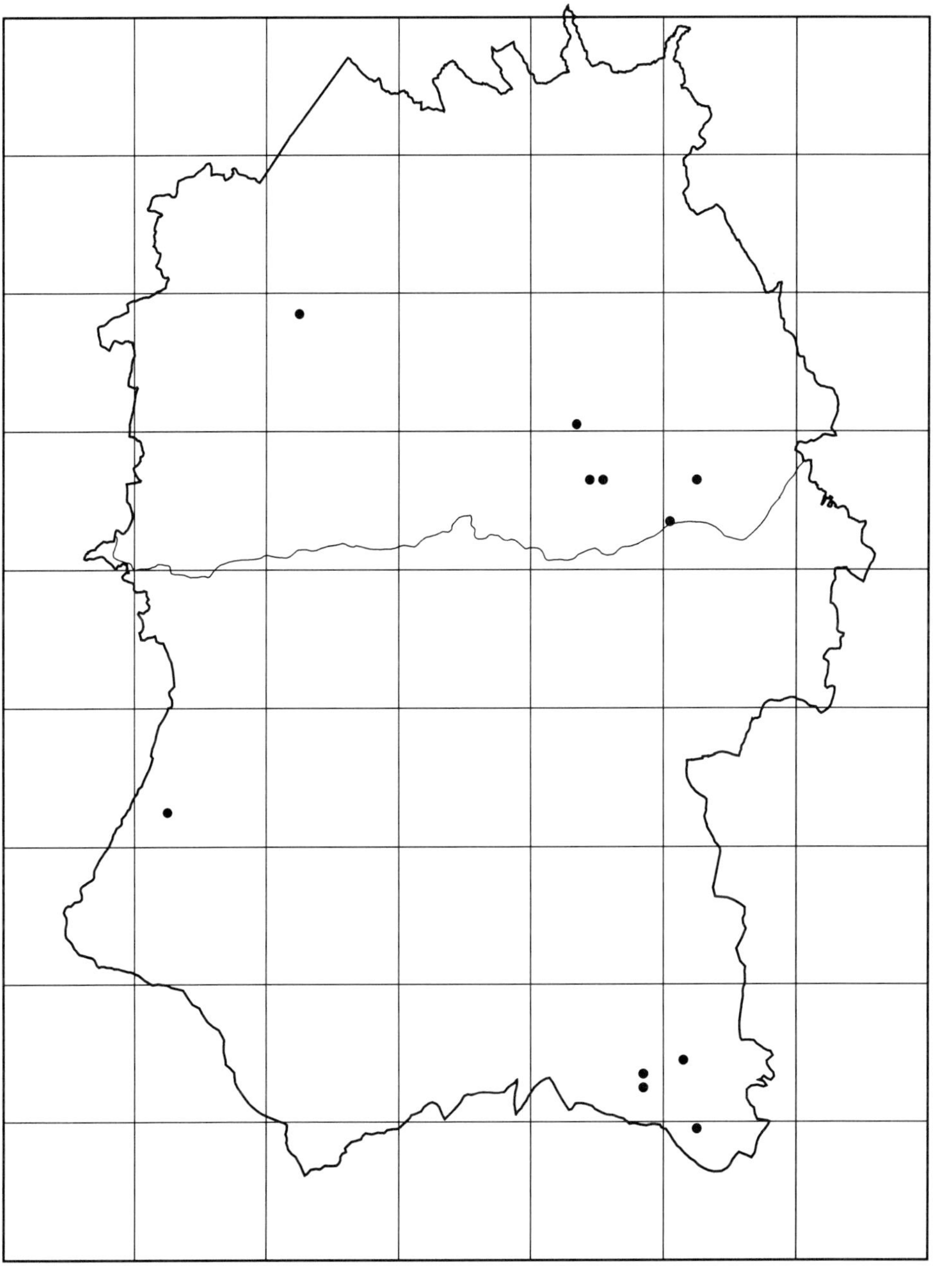

Figure 37.1. Red deer: distribution

Order Artiodactyla — Family Cervidae

Sika deer — *Cervus nippon*

Intermediate in size between red and roe deer with a lightly spotted coat which is reddish-brown in summer. Maximum shoulder height 85cm.

Handbook, 504-508; *Atlas*, 130-131; *Review*, 97-98.

Status

Introduced. Common in places.

The total pre-breeding population in Britain is estimated at 11,500 and may be slowly increasing locally.

Nationally

Widespread but scattered distribution in England and Scotland. All populations have originated from animals which have escaped from parks or animals which have been deliberately released.

Wiltshire

No special survey of the Sika deer has been published.

First recorded in the Salisbury district in 1959 (*SDNHS Report* 1959); further records 1964-5 and 1974.

Four records of which three have been used in this analysis. All but one of the records are from south of Salisbury, near the borders of the New Forest, where the original sightings, 1959-1974, were made. The other record is of an animal crossing a road at Durrington, north of Salisbury in 1977 (fig. 38.1).

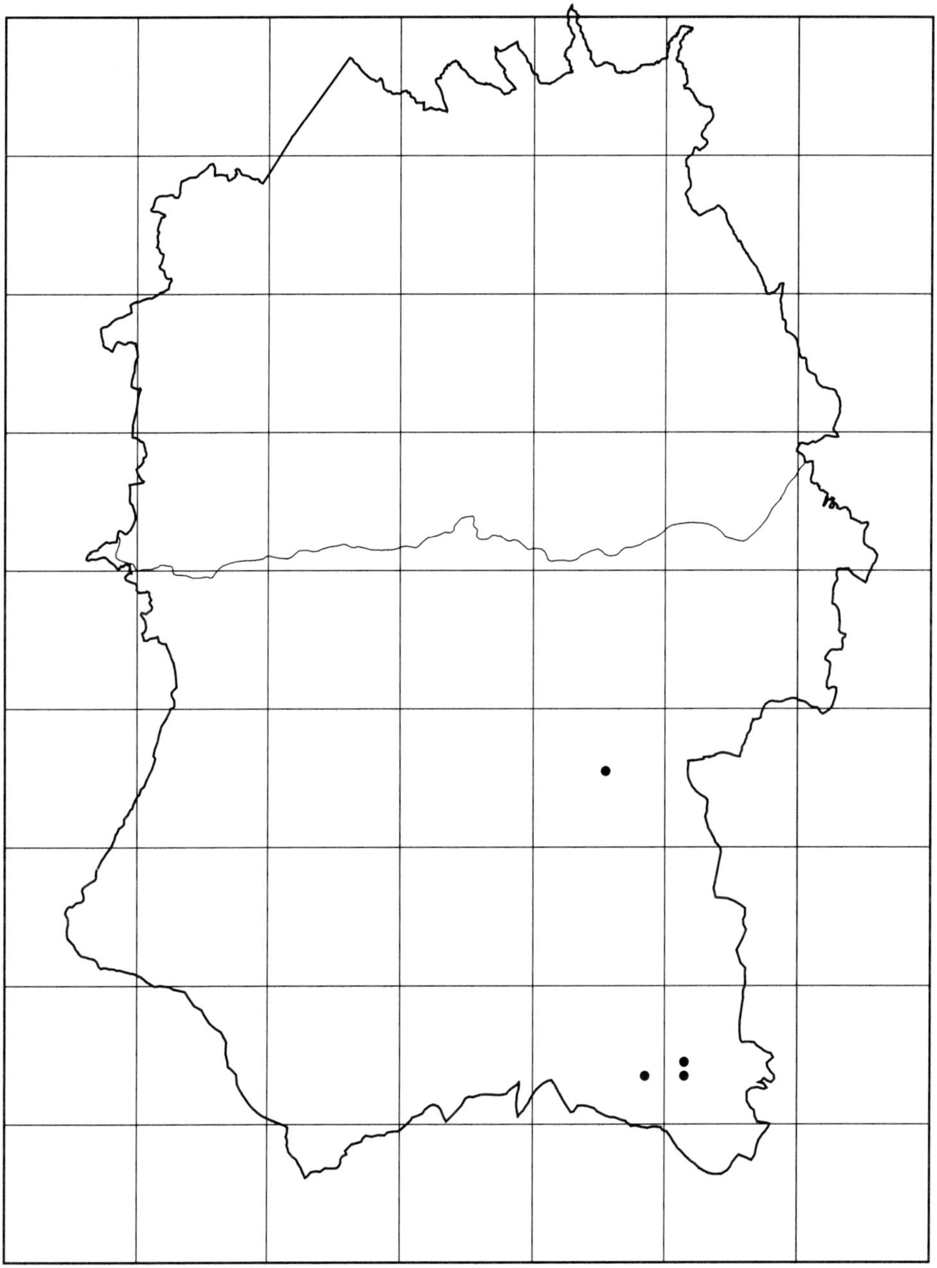

Fig. 38.1. Sika deer: distribution

Order Artiodactyla — Family Cervidae
Fallow deer — *Dama dama*

Medium-sized deer with a coat which shows a wide variation in colour but which typically, in summer, is fawn with white spots. Maximum shoulder height 100 cm.

Handbook, 509-518; *Atlas*, 132-133; *Review*, 98-99.

Status

Introduced. Common where it occurs.

The total pre-breeding population in Britain is estimated at 100,000 and may be slowly increasing.

Nationally

Widely distributed in England, Wales and Scotland but the main centres of population are in southern and central England. Fallow deer are thought to have been introduced by the Normans in the 11th century. Present populations are descended from animals released directly into forests or animals which have escaped from deer parks. The favoured habitat of the species is deciduous woodland.

Wiltshire

No special survey of the fallow deer has been published.

Fallow deer in Cranborne Chase, Grovely Forest and Clarendon Park were mentioned by Aubrey (c.1660). These, and populations derived from the herd in Savernake Forest, account for most of the Wiltshire records.

99 records of which 66 have been used in this analysis. 77% of records are of live animals (fig. 39.1). Distribution is centred on Cranborne Chase, Grovely Forest and Clarendon Park with some isolated populations or records of single animals in the west of the county, possibly from herds from neighbouring counties (fig. 39.2). Of the 52 records of habitat, 42 are of woodland with small numbers of records for downland, parkland, farmland and road/railbank. Of 14 records of field signs, 12 are of prints, one is of an animal 'calling' and the other is unspecified. There is one record of an animal dead on a road.

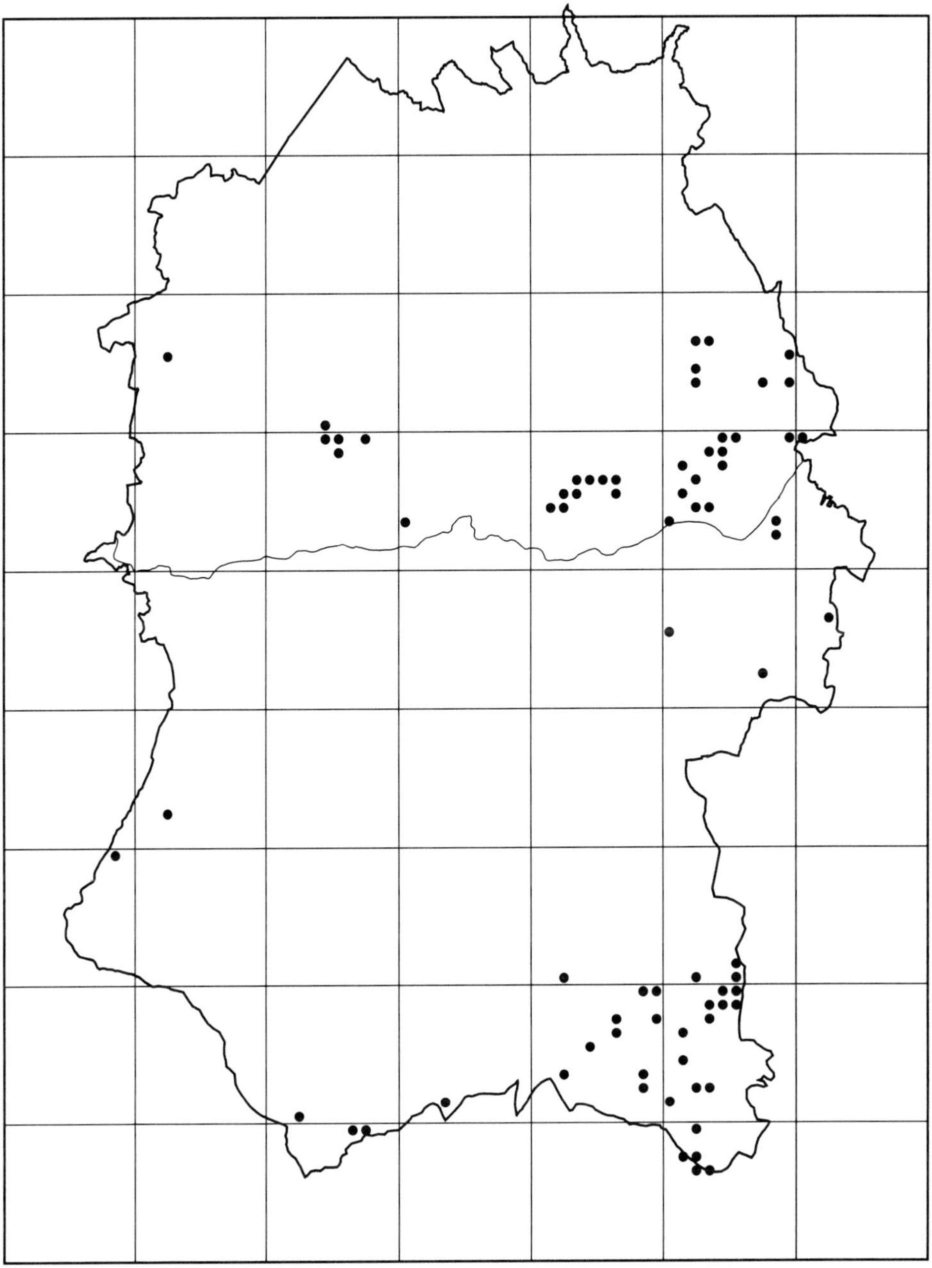

Fig. 39.2. Fallow deer: distribution

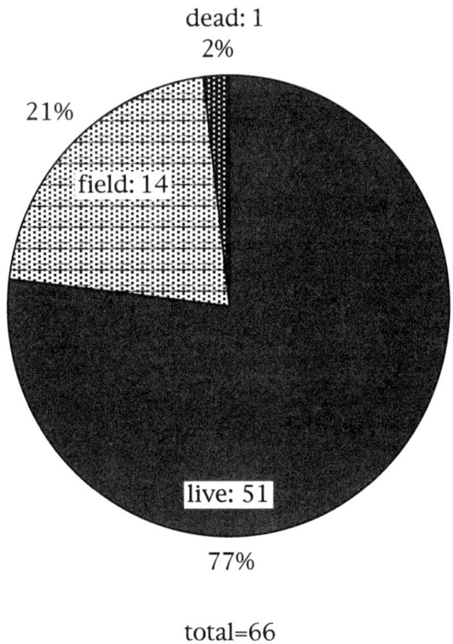

Fig. 39.1. Fallow deer: number and nature of record

Order Artiodactyla — Family Cervidae

Roe deer — *Capreolus capreolus*

Small to medium-sized deer with a short muzzle and distinctive black nose and very short tail with pale rump patch. Maximum shoulder height 75 cm.

Handbook, 518-526; *Atlas,* 134-135; *Review,* 100-101.

Status

Native.

The total pre-breeding population in Britain is estimated at 500,000 and is increasing. The species is also extending its range.

Nationally

There is a native population widely distributed in northern England and Scotland. Introductions from Scotland and Europe have given rise to a widely distributed population in southern England. There are scattered records from Wales and central England. Roe deer are generally associated with mixed deciduous and coniferous woodland.

Wiltshire

No special survey of the roe deer has been published.

The roe deer became extinct in southern England in historic times; bones found amongst rubbish dating from the thirteenth century in Cranborne Chase are possibly from local animals (Pitt Rivers 1890). The first recorded re-introduction was to Blackmore Vale, Dorset in 1820 (Hony 1917) and these animals are thought to have founded populations which have subsequently colonised parts of Wiltshire.

The first recent Wiltshire records are c.1900-1910, from Great Ridge Wood, Hindon (Hudson 1910) and Dinton, 1909 (Hony undated). Roe deer were first reported from Savernake Forest in the 1940s (*MCNHS Report* 1945). Records suggest an expanding population in south Wiltshire in the 1950s and 1960s (*SDNHS Reports*).

698 records of which 615 have been used in this analysis. 85% of records of live animals (fig. 40.1) Widely distributed and present in most places where there is suitable habitat (fig. 40.2). Recorded in a range of habitats but predominantly woodland (fig. 40.3). It is not clear to what extent the non-woodland records reflect a break with the usual habitat or represent animals on the move. Of 72 records based on field signs, 54 are of prints, eight are of droppings, three are of animals calling and three are of skeletal remains. Of 17 records of mortality, 10 are of animals dead on roads, five are of casual finds and two are of animals killed by dogs.

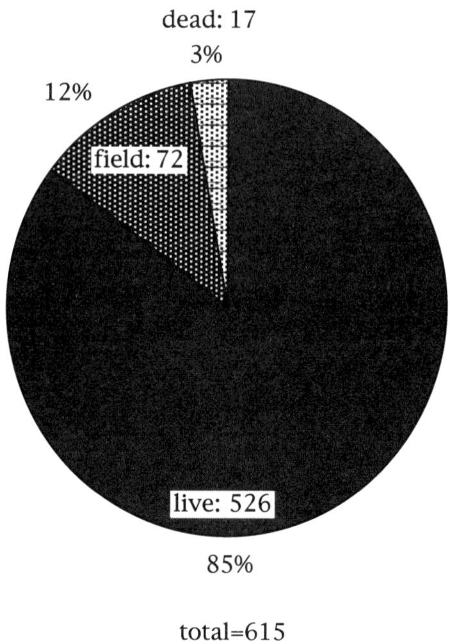

Fig. 40.1. Roe deer: number and nature of record

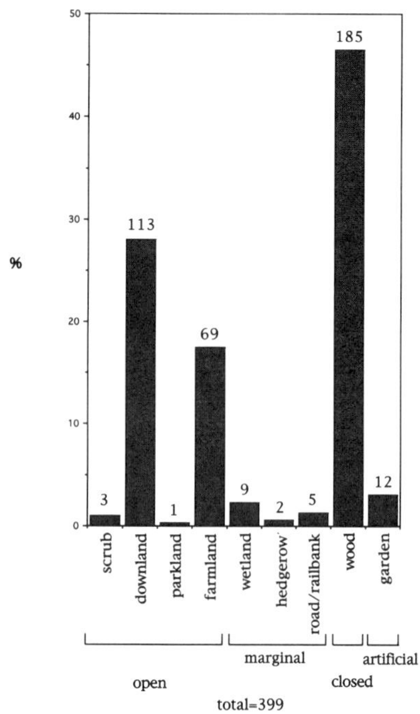

Fig. 40.3. Roe deer: habitats

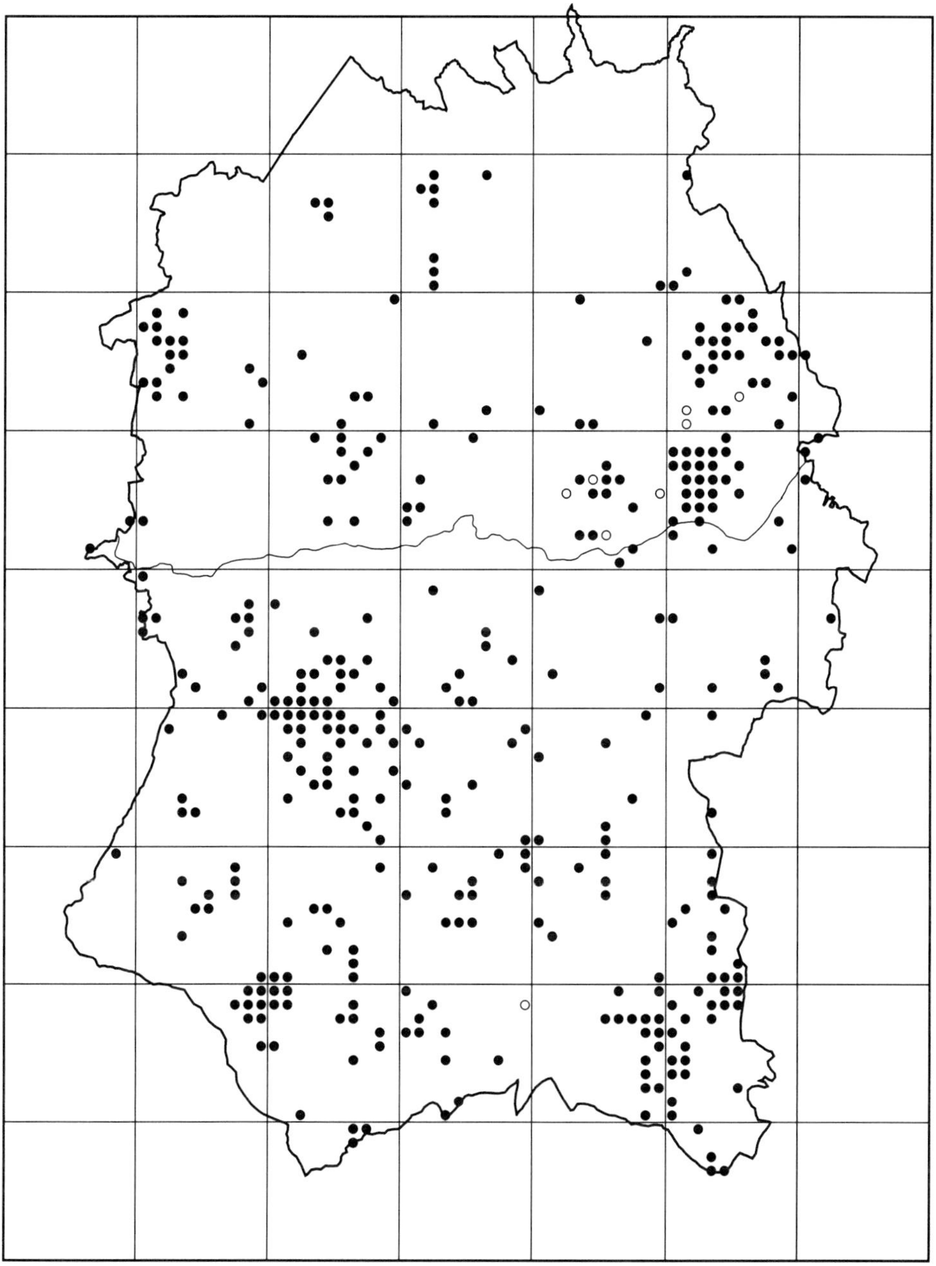

Fig. 40.2. Roe deer: distribution

Order Artiodactyla — Family Cervidae

Muntjac — *Muntiacus reevesi*

Small deer with brown coat and distinctive white underside to the long tail. Maximum shoulder height 50 cm.

Handbook, 526-532; *Atlas*, 136-137; *Review*, 102-104.

Status

Introduced. Common where it occurs.

The total pre-breeding population in Britain is estimated at 40,000 and is increasing rapidly.

Nationally

The British population originates from introductions to Woburn Park in the early twentieth century. Muntjac are now widespread in central and southern England and present in parts of Wales. The species is usually associated with habitats offering dense cover.

Wiltshire

No special survey of the muntjac has been published.

The first record for Wiltshire was of an animal shot in Savernake Forest in 1938 (Chapman, Harris and Stanford 1994). Animals were reported from Witherington Wood in 1958 and Froxfield c.1966 (Chapman, Harris and Stanford 1994) which may have originated from the Savernake introduction. An animal was reported from Standlynch, south of Salisbury in 1968 (*SDNHS Report* 1968) and this appears to be the earliest sighting on the fringe of the New Forest.

Chapman, Harris and Stanford (1994) suggest that the records from the three estates, Bowden, Bowood and Spye Park dating from 1966 represent a population having its origin with a separate introduction.

53 records of which 40 have been used in this analysis. 70% of records of live animals (fig. 41.1). Wide but scattered distribution. Present in most places where there is suitable habitat (fig. 41.2). Recorded in a range of habitats but predominantly woodland. Of 11 records of field signs, eight are of prints, one is of droppings, and two are unspecified. The one record of mortality is of an animal dead on a road.

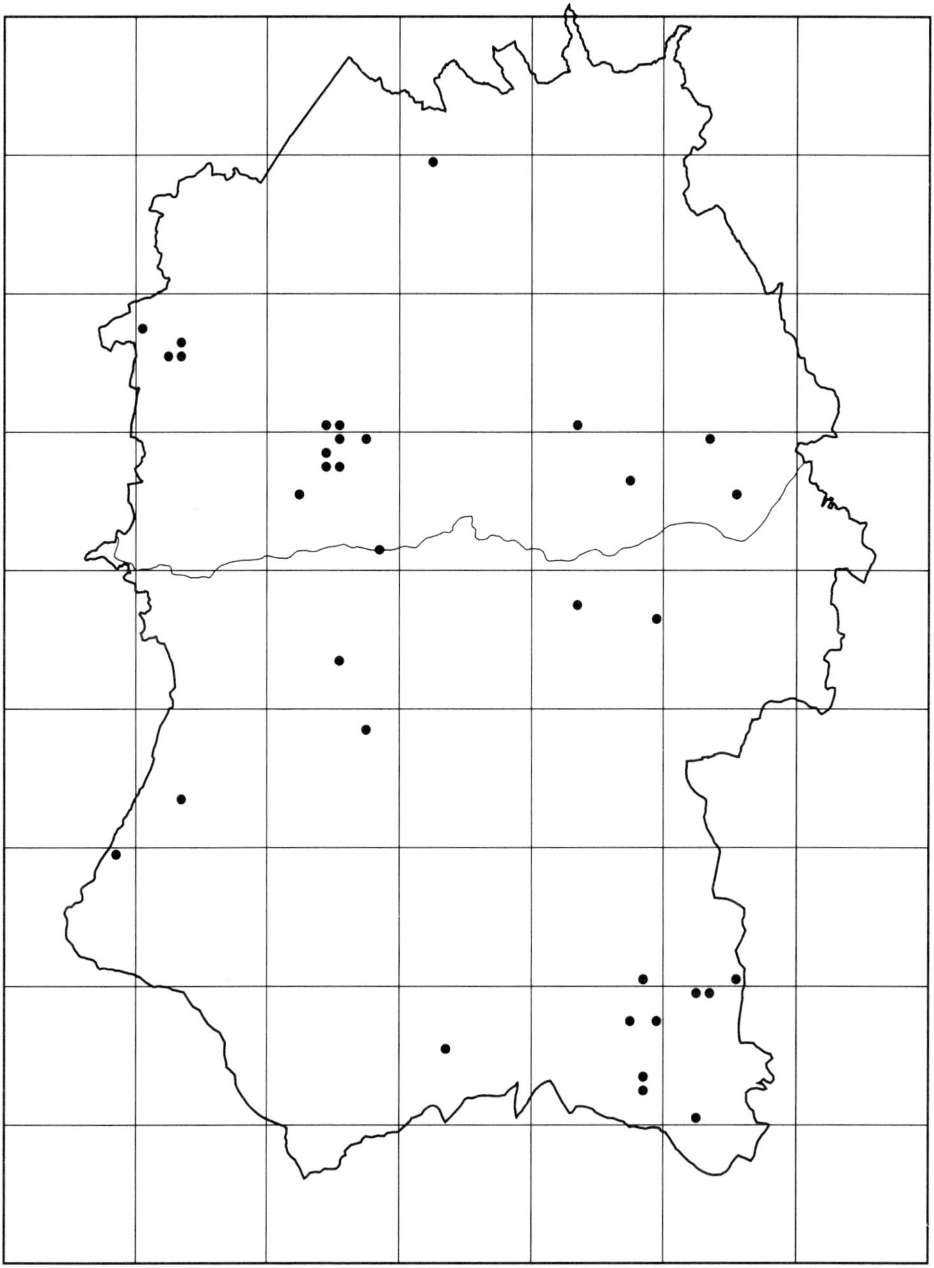

Fig. 41.2. Muntjac: distribution

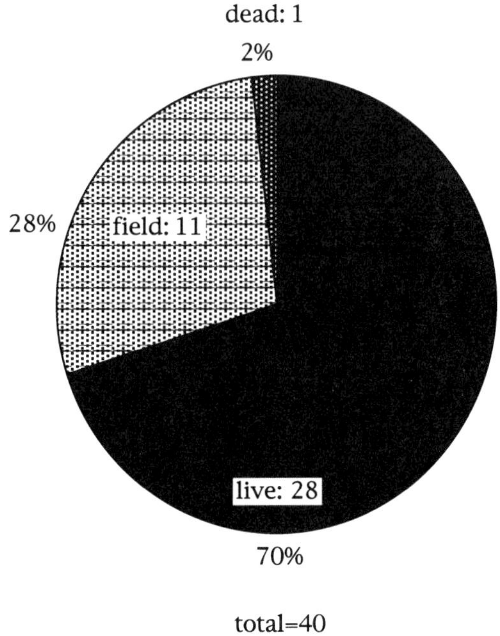

Fig. 41.1. Muntjac: number and nature of record

REFERENCES

Arnold, H. 1993. *Atlas of Mammals of Britain.* London, HMSO.

Atkinson, R.P.D., Macdonald, D.W. and Johnson, P.J. 1994. The status of the European mole *Talpa europaea* L. as an agricultural pest and its management. *Mammal Review,* 24, 73-90.

Aubrey, J. c.1660. *Natural History of Wiltshire.* Newton Abbot, David and Charles, 1969 reprint.

Bell, T. 1837. *A History of British Quadrupeds.* London, John van Voorst.

Benham, E. 1953. The distribution of squirrels in Dorset. *Proceedings of the Dorset Natural History and Archaeological Society,* 74, 121-132.

Bright, P. 1993. Habitat fragmentation – problems and predictions for British mammals. *Mammal Review,* 23, 101-111.

Bright, P.W. 1996. Status and woodland requirements of the dormouse in England. *English Nature Research Report No. 166.*

Browne, M. 1977-1988. Mammal Reports. *Wilts. Nat. Hist. Mag.* 72, 30-39; 73, 32-40; 74, 38-41; 75, 34-40; *WANHS Annual Report,* 1981, 39-44; 1982, 57-61; 1983, xxii-xxvii; 1984, xxii-xxviii; 1985, xxvii-xxxiii; 1986, xxx-xxxvii; 1987, xxix-xxxv; *WANHS Natural History County Recorders' Reports,* 1988.

Browne, M. 1983-1990. *WAM,* 77, 123-137 (water vole); 79, 206-213 (dormouse); 80, 197-210 (shrews); 81, 111-122 (hedgehog); 82, 147-155 (mole); 83, 190-200 (common rat).

Browne, M. and Gillam, B. 1974. The harvest mouse (*Micromys minutus*) in Wiltshire. *WAM,* 69, Part A, 47-48.

Bunce, R.G.H., Barr, C.J, Clarke, R.T., Howard, D.C. and Lane, A.M.J. 1996. Land classification for strategic ecological survey. *Journal of Environmental Management,* 47, 37-60.

Chafin, W. 1818. *Anecdotes and History of Cranborne Chase.* London, J. Nichols.

Chapman, N., Harris, S. and Stanford, A. 1994. Reeves' Muntjac *Muntiacus reevesi* in Britain: their history, spread, habitat selection, and the role of human intervention in accelerating their dispersal. *Mammal Review,* 24, 113-160.

Clay, R.C.C. 1928. Some notes on the history of Fovant. Unpublished typescript, WANHS.

Corbet, G.B. and Harris, S. 1991. *The Handbook of British Mammals* (third edition). Oxford, Blackwell.

Cresswell, P., Harris, S. and Jefferies, D.J. 1990. *The history, distribution, status and habitat requirements of the badger in Britain.* Peterborough, Nature Conservancy Council.

Dillon, P.J. 1977. E.H. Goddard's natural history notes, 1873-1887. *Wilts. Nat. Hist. Mag.* 72, 3-9.

Dillon, P. 1977. Some recent Wiltshire owl pellet analyses. *Wilts. Nat. Hist. Mag.* 72, 32-33.

Dillon, P.J. 1984. Natural history manuscripts in Devizes Museum. *WAM,* 78, 105-113.

Dillon, P.J. 1990. An historical perspective on the occurrence of *Rattus rattus* and *R. norvegicus* in Wiltshire. *WAM,* 83, 215-217.

Dillon, P. and Browne, M. 1975. Habitat selection and nest ecology of the harvest mouse, *Micromys minutus* (Pallas). *Wilts. Nat. Hist. Mag.* 70, 3-9.

Dillon, P. and Browne, M. 1984. Small rodents in Wiltshire: voles and mice. *WAM,* 78, 94-104.

Dillon, P. and Jones, E.L. 1986. Trevor Falla's vermin transcripts for Devon. *Devon Historian,* 16, 3-8.

Dillon, P.J. and Noad, P. 1980. A catalogue of the Natural History Library of the Wiltshire Archaeological & Natural History Society, Part 1 printed works. *Wilts. Nat. Hist. Mag.,* 75, 2-19.

Dunthorn, A.A. and Errington, F.P. 1964. Road casualties among mammals, reptiles and amphibians in Wiltshire. *WAM,* 59, 207-210.

Gillam, B. 1967. The distribution of badgers in Wiltshire, 1966. *WAM,* 62, 145-152.

Gillam, B. 1973. Amphibia, reptile and mammal notes for 1970-1972. *WAM,* 68, Part A, 7-10.

Gillam, B. 1993. *The Wiltshire Flora.* Newbury, Pisces Publications.

Goddard, E.H. 1942. Natural History Notes, 1873-1887. MS notebook, WANHS. (Edited version Dillon 1977).

Grose, D. 1957. *The Flora of Wiltshire.* Devizes, Wiltshire Archaeological and Natural History Society.

Harris, S., Morris, P., Wray, S. and Yalden, D. 1995. *A Review of British Mammals.* Peterborough, Joint Nature Conservation Committee.

Harting, J.E. 1880. *British Animals Extinct Within Historic Times.* London, Trubner and Co.

Hawksley, J.C. 1978. A colony of brown rats. *WANHS Natural History Bulletin,* 10, 13-14.

Hawksley, J. 1980. A coypu in the River Kennet. *WANHS Natural History Bulletin,* 13, 5-6.

Hony, G.B. 1917. The mammals of Wiltshire. *WAM,* 39, 15-21.

Hony, G.B. undated. Natural History Notes. Unpublished manuscript in the Library of WANHS.

Hudson, W.H. 1910. *A Shepherd's Life*. London, Methuen.

Jackson, J.E. 1875. Wulfhall and the Seymours. *WAM,* 15, 140-207.

King, N.E. and Sheail, J. 1970. The old rabbit warren on Fyfield Down, near Marlborough. *WAM,* 65, Part A, 1-6.

Maton, G. 1843. *The Natural History of a part of the County of Wilts comprehended within the distance of ten miles around the City of Salisbury.* London, J.B. Nicholls & Son.

Meyrick, E. 1896. List of vertebrate animals of the Marlborough District. *Report of the Marlborough College Natural History Society,* 38-42.

Millais, J.G. 1904-6. *Mammals of Great Britain and Ireland.* London, Longman and Green.

Montagu, G. c.1785. Natural History of Wiltshire. Unpublished manuscript in the Linnean Society of London.

Montagu, G. 1803. Notebook of the Late Geo. Montagu, F.L.S. Unpublished manuscript in the Linnean Society of London.

Morris, P.A. 1993a. *The Red Data Book for British Mammals.* London, The Mammal Society.

Morris, P.A. 1993b. British mammals – their status, research needs and likely future. *Mammal Review,* 23, 167-176.

Peirson, L.G. 1923. Notes on the mammals, reptiles and amphibians of the Marlborough district. *Report of the Marlborough College Natural History Society,* 72, 42-44.

Pitt Rivers, A.H. Lane-Fox. 1890. *King John's House, Tollard Royal.* London, Printed Privately.

Reason, P., Harris, S. and Cresswell, P. 1993. Estimating the impact of past persecution and habitat changes on the numbers of badgers *Meles meles* in Britain. *Mammal Review,* 23, 1-15.

Scott-White, S. 1991. Badger casualties on Wiltshire roads. *WANHS Newsletter,* 24, 5.

Scrope, G.P. 1852. *History of the Manor and Ancient Barony of Castle Combe in the County of Wilts.* Printed for private circulation. Copy in the Library of WANHS.

Smith, A.C. 1875. A plea for the moles. *WAM,* 15, 308-320.

Smith, A.C. 1879. A sketch of the parish of Yatesbury. *WAM,* 19, 319-359.

Straton, C.R. 1902. An English Manor in the time of Elizabeth. *WAM,* 32, 288-310.

Tapper, S. 1992. *Game Heritage. An Ecological Review from Shooting and Gamekeeping Records.* Fordingbridge, Game Conservancy Ltd.

Ward, J.M. 1981. What the cat brought in. *WANHS Natural History Bulletin,*15, 6-7.

Williams, A. 1922. *Round About the Upper Thames.* London, Duckworth and Co.

Wordsworth, C. 1907. Customs of Wishford and Barford in Grovely Forest. *WAM,* 35, 283-316.

Appendix 1
LIST OF RECORDERS OF MAMMALS IN WILTSHIRE

B.H. Adams
Mr. & Mrs. P. Adams
H.A. Adderley, Warminster
N. Ajax-Lewis
Mrs. G.P. Alford, Salisbury
Mr. D. Allen
Miss J. Allen, Calne
Mr. N. Allen, Chippenham
Mr. N. Anderson, Salisbury
Sir C. Andrewes, Salisbury
Mrs. H. Antrobus, Devizes
Miss M. Archer, Chippenham
A.J.A. Arengo-Jones, Swindon
Miss M.E. Ashbery, Devizes
Countess J. Badeni, Malmesbury
D. Bailey, Salisbury
S.G. Bailey
Mr. A.R. Baker, Calne
Mrs. J.E. Baker, Westbury
D.E. Balar
Mrs. P.M.G. Baldry, Malmesbury
Mr. T.C. Bale, Salisbury
Mrs. L. Balfe, Westbury
Mrs. R.G. Barnes, Chippenham
Mrs. A.E. Barnitt, Tetbury
P.D. Barnitt, Tetbury
R.S. Barron
M. Bastable, Trowbridge
E. Batchelor, Salisbury
Mr. A.R. Beazer, Chippenham
Mr. P.A. Beazer, Chippenham
Mr. & Mrs. P.J. Beazer, Devizes
Mr. F.M.E. Beeley, Salisbury
Mr. D.A. Beeson
Mrs. D. Bell, Chippenham
R. Bennet, Westbury
Mr. H.F. Bennett, Chippenham
C. Bindon
P.F. Bird, Bristol
Miss W. Bishop, Westbury
Mr. & Mrs. F.C. Booth
D.M. Bracher, Westbury
K.J. Bradley, Warminster
Mr. N.B. Brimble, Corsham
Bristol & District Amalgamated Anglers
Bristol Naturalists Society
British Field Sports Society
British Museum (Natural History)
Broad Town School
Major Gen. J.M. Brockbank, Salisbury
Mrs. B. Brooke, Devizes
J. Brookhouse, Devizes
Mr. D.J. Brotheridge, Swindon
Mr. E. Brown, Salisbury
J.R. Brown, Salisbury
Mrs. M. Browne, Chippenham
Mr. R.B. Browne, Chippenham
Mr. S.H.B. Browne
Mr. G.P. Bryant, Calne
A.D. Bullifent, Salisbury
Miss C. Burke, Devizes
Mrs. P. Burnand, Tetbury
Mr. K.B. Burry, Devizes
J.A. Burton
Miss J. Burton, Trowbridge
Mrs. G. Butt, Chippenham
Mr. E.J.M. Buxton, Chippenham
Major & Mrs. C.W. Byham, Trowbridge
Miss M. Campbell, Trowbridge
Mrs. S. Campbell-Ross, Pentridge
Mr. R. Capel
Mrs. J.M. Capjon, Marlborough
Miss A.M. Cardus, Malmesbury
J.A. Carman, Salisbury
B. Carnell
B.D. Castle, Marlborough
Mr. P.E. Castle, Swindon
Miss V.M. Caswall, Devizes
R.N.L. Chadwick
Mrs. J.L. Channing, Devizes
Mr. J. Charnaud, Chippenham
Mr. T. Chivers, Chippenham
Mr. H. Chivers, Calne
Mrs. Z. Chivers, Chippenham

M.H. Clarke
M.L. Clarke, Salisbury
Lt. Col. C.H.F. Coaker, Marlborough
Miss W.G. Cobbett
Miss A.M. Coburn
Mrs. J.A. Cochran
Mrs. P.C.U. Cogdell, Marlborough
Mrs. J.E. Colley, Devizes
Mr. M. Collins
W.J. Collins, Warminster
Mrs. M.E. Compton, Chippenham
Mrs. A. Corbett, Devizes
R. Cowdrey, Chippenham
Mrs. E. Cowley, Chippenham
Mr. R. Cowlin, Epping
D.F. Cox
Mr. P.A. Cox, Warminster
Miss S. Creane, Warminster
R.M. Cross
Mrs. S.P. Cross, Pewsey
Mr. M. Curtis, Swindon
M.J. D'Oyly, Stroud
Daily Express
Mr. J. d'Arcy, Westbury
Dauntsey Fauna List
Dauntsey School, Devizes
Mrs. J. Davies, Salisbury
Mr. M.J. Davies, Salisbury
Mr. R. Daw, Marlborough
E.C. Deadman, Calne
Mrs. D. Dean, Warminster
N.G. de Fossard, Chippenham
G. Delamain
Mrs. E. Errick, Chippenham
P. Derrick
Mrs. J.E. Dillon
Dr. P.J. Dillon
E. Dixon
D.G. Dolman, Oxford
A.E. Downer, Salisbury
Mr. G. Drake, Devizes
Mr. M.T. Drew, Chippenham
A.A. Dunthorn
Mr. P. Durnford
Miss A.E. Earl, London
K.Y. Eccles
I. Edelman, Avebury
H. Edmunds, Salisbury
Mrs. R. Edwards, Chippenham
Mr. S.B. Edwards, Swindon
Mr. L. Edwins
Captain H. Ennion, Marlborough
F.P. Errington
Mr. E.J. Escott, Chippenham
J.D. Ethelston
I.M. Evans
L. Excell, Swindon
Mr. E. Farquhar, Chippenham
Lt. Col. Sir P. Farquhar, Chippenham
Mr. A.S.R. Featherstone, Salisbury
Mrs. B. Fergusson, Salisbury
D. Fielding
Mr. R.F. File, Devizes
Mrs. A.O.J. Fitzmaurice, Pewsey
Miss D. Forbes, Salisbury
Miss K.G. Forbes, Warminster
G.H. Forster, Farnham
Mr. & Mrs. G.M. Foxwell, Corsham
J. French, Marlborough
Mr. M. Fuller, Bradford-on-Avon
Mr. R.D. Fussell, Devizes
Mrs. J.E. Gaffney
Game Research Association, Fordingbridge
Mrs. H.M. Gardner, Westbury
Mrs. S.M. Gibbon, Marlborough
Miss B. Gillam, Devizes
B. Gillett, Marlborough
C. Gingell, Chippenham
Mr. D. Glue
E.H. Goddard
Mr. F.R. Gomme, Chippenham
Mrs. C. Goodfellow, Chippenham
Mr. M.J. Gould, Devizes
J.R. Govett, Trowbridge
Mrs. T.A.K. Gow, Salisbury
Mr. & Mrs. R. & V. Gowland, Chippenham
Mrs. D. Graiff, Salisbury
Mrs. M.E.E. Gray, Warminster

I.J. Gray, Swindon
Miss Y. Greasley, Trowbridge
R.G.C. Green, Trowbridge
Mrs. V. Greenaway, Swindon
R.A. Greenway, Swindon
Dr.I. Grier, Westbury
Mr. R. Griffiths, Swindon
P. Grist
Mr. D. Grose
Mrs. I. Grose
P. Gwyn
Mr. P.R. Hamilton-Leggett, Bradford-on-Avon
P. Harding, Marlborough
T. Hardy, Bath
Dr. S. Harris, Bristol
Revd. D.P. Harrison
T.D. Harrison, Reading
J. C. Hawkesley, Marlborough
R. Hayden
B.J.W. Heath, Marlborough
Mrs. J. Heath
Mr. M. Heath
M. Heath, Pewsey
Miss B. Herrod-Taylor, Devizes
Mrs. D. Herrod-Taylor, Devizes
I.R.S. Heslop
Mrs. C.S. Hett
Miss E. Higgins, Salisbury
J.B. Hindley, Salisbury
Mr. A. Hodges, Corsham
Mrs. D.M. Holdsworth, Marlborough
Mrs. A.R. Holeton
Mr. B. Hollies, Devizes
G.B. Hony
Mrs. N. Hooke, Devizes
B.W. Hookins, Shaftesbury
H.N. Hooper
Mr. D. Horton
P.J. Horton
Mr. J. Hounslow
J. Howard
Mr. G.A. Howe, Warminster
Mr. J. Howitt, Westbury
Mrs. M.I. Hoyle, Marlborough
Mr. K.A. Huggett, Westbury
Miss G. Huggins, Bath
H.C. Hulbert, Corsham
Mr. D.G. Humphries, Melksham
Mr. K. Humphry, Chippenham
F. Humphrys, Swindon
Mr. R. Hunt
Mrs. E. Hunter, Marlborough
Miss E. Hurrell
Mr. D. James
Mrs. G. James, Chippenham
R.T. James
A. Jenkins
Mrs. K. Jenkins, Chippenham
Mrs. J. Jenkyns, Salisbury
Mr. C. Jennings
Mr. V. Jennings, Devizes
J.A. Johnson, Chippenham
Mrs. M.M. Jones, Trowbridge
Miss O.R. Jones, Westbury
Mrs. J. Jordan, Marlborough
Miss J.M. Jordan, Salisbury
Mr. P. Kane
Prof. H.E.M. Kay, Pewsey
Miss J. Kedge, Warminster
Mrs. A. Kelham
Mr. J. Kendall, Chippenham
Lady E. Kennet, Marlborough
Miss J. King, Devizes
Mr. N.E. King
D.W. Kingston, Swindon
Mr. O.A.W. Kite, Netheravon
Mr. W. Knight
J.W. Lambert, Melksham
Mr. S. Lane, Devizes
Mrs. G.A. Langdon, Bradford -on- Avon
Mrs. B. Last
R.W. Lemon, Salisbury
Miss E.J. Lenton
Mrs. G.M. Lewis, Warminster
Miss G. Lloyd, Chippenham
Mrs. S. Long, Devizes
Mr. & Mrs. D. Lovelock, Calne
J. MacFadyen, Marlborough
I.G. Mace, Malmesbury

Ft. Lieut. J.E. Major, Shaftesbury
K. Mantock
Marlborough College Natural History Society
Mrs. J.A. Martin, Chippenham
Miss P. Martin, Westbury
R.A. May, Swindon
Mr. R. Maycock, Salisbury
Mr. N. McCarter
K. McDowell, Westbury
Mrs. D.G. Mead, Salisbury
Mrs. J. Melsom, Malmesbury
Mr. B. Merry
E. Meyrick
Mr. T. Miles, Chippenham
Ministry of Agriculture Food and Fisheries
Mrs. M.S. Mitchell, Salisbury
Mrs. A. Moffat, Malmesbury
Mrs. J. Moffat, Calne
A. Money-Kyrle
R. Moody, Trowbridge
Dr. P. Morris
Miss. J.M. Morrison, Westbury
Mr. & Mrs. J.B. Muns, Marlborough
National Game Census
Nature Conservancy Council, Devizes
R.S. Newall
Mr. P.E. Newbery
Mr. H.C. Newton, Marlborough
Mrs. J. Newton, Marlborough
Mr. P.N. Newton, Marlborough
Mr. R. Newton, Marlborough
Mrs. K. Nicol, Pewsey
M. Norgate, Trowbridge
Cdr. A.Y. Norris, Fareham
J.S. Notman, Salisbury
Major Oates
Miss E.L.M. Occomore, Salisbury
Dr. J.E. Oliver, Marlborough
Mrs. C.A. Ollis, Chippenham
Miss E.D. Overend, Chippenham
Mr. C. Owen
G.S. Oxford
Mr. D. Painter, Marlborough
M.K. Palfreman
Mr. D. Palmer, Chippenham
Mr. S.M. Palmer
W.F. Paker
Mrs. B. Pattrick
Mr. D.B. Paynter
W.A.O. Pearce, Chippenham
Mr. R.E. Penny, Trowbridge
Mr. E. Pettit, Salisbury
Mr. P. Phillipson, Buxton
Mrs. S.A. Phillipson, Buxton
Mr. C.M.R. Pitman, Salisbury
Mr. D.J. Pollard, Melksham
E.W. Powell
J.T. Price
P. Racey
R. Raddon, Devizes
Lady Radnor
Mr. R.D. Ransome
M. Rebane
F.G. Reeman
Mrs. J.M. Richards, Devizes
The Hon. Mrs. H.C. Ridley, Warminster
Mr. N.F. Rigiani, Salisbury
D.W. Roberts
E.E. Roberts
H.D. Roberts, Bath
R. Roberts, Devizes
D. Robinson, Devizes
Dr. P. Robinson
Mr. J. Rogers
Mr. J.C. Rolls, Melksham
Miss S.F. Rooke, Salisbury
G. Rowles, Devizes
R.G. Rugg
Mr. M. Russell
C.D. Ryan, Salisbury
Mr. A.J. Rycroft, Pewsey
Mr. N. Sadler, Chippenham
Salisbury & District Field
Salisbury & District Natural History Society
R.E. Sandell
Mr. P.A. Sandison, Salisbury
Mr. D. Saunders, Haverfordwest

P.R. Saunders, Salisbury
Mrs. A. Sawyer, Warminster
Lt. Col. E.L. Sawyer, Warminster
F.M. Sawyers, Warminster
Mrs. S. Scamell, Salisbury
L. Scoging, Marlborough
Mr. P.R. Scott, Chippenham
G. Scriven, Salisbury
R.C. Seal, Melksham
Mr. J. Sellar
M.H. Shannon, Calne
Mr. W.D. Shaw, Westbury
R.C.H. Shepherd
T. Sherwen, Yeovil
Dr. D. Shirt, Marlborough
Mrs. H. Shortt, Salisbury
S.B. Sidebottom, Salisbury
B. Siggers, Westbury
Mr. A. Smith, Trowbridge
E.G. Smith, Trowbridge
D.J. Smith, Bath
Mr. E. Smith, Marlborough
Mrs. H. Smith, Corsham
Mr. P.A. Smith, Corsham
Mr. P. Smith, Corsham
Mrs. H. Snipp, Newbury
C. Spray
A. Squires, Andover
Dr. R.E. Stebbings
Dr. E. Steel, Cambridge
Mr. & Mrs. E. Stephens, Bradford-on-Avon
Miss B.M. Stevens, Bradford-on-Avon
Mr. J.A. Stevenson
W. Stevenson
Miss M.J. Stokes, Warminster
Mr. G. Stone
Mr. A. Stonell, Devizes
Mrs. J. Stopford Beale, Chippenham
B. Stratford, Salisbury
Miss C.A. Street, Bradford-on-Avon
Mrs. A. Summers, Pewsey
Mrs. J. Swanborough, Chippenham
Lt. Col. D.C. Swift, Salisbury
B.H.C. Sykes, Devizes
Mr. R.G. Symes
F.G. Tanner, Warminster
Mr. P. Tasker
P.J. Taton-Brown
D. Taylor
Miss M.K. Taylor, Marlborough
Mr. & Mrs. W.G. Teagle, Swanage
Miss V. Temple, Salisbury
Mr. P. Thomas, Calne
Mrs. S.M. Thomson, Salisbury
Mrs. S. Thomson, Swindon
Mrs. M. Thorne
Mr. & Mrs. R. Thornton, Marlborough
C.B. Ticehurst
P.C. Tinning, London
P. Titt
R. Titt
A.M. Tittnesor
Mr. & Mrs. F. Tolfree, Chippenham
R.S. Tranchant
Mr. G.J. Troughton, Salisbury
Mrs. P. Turnbull
Mr. R. Turner, Westbury
Miss. P.M. Twiney, Warminster
Mr. M. Tyler, Swindon
Mrs. S.J. Tyler
D. Waddilove
L.D. Walker, Marlborough
C. Ward
Miss J.M. Ward, Bradford-on-Avon
Mrs. P.A. Ward, Pewsey
R. Ward, Chippenham
Mr. P. Waring, Oxford
B.C.H. Warren, Salisbury
J.M. Waters
A. Watson
Miss D.M. Wear, Salisbury
J.I. Webb, Chippenham
G.L. Webber
Mr. A. Weeks, Chippenham
Mr. P. Weeks, Chippenham
Mrs. H.W. Welch, Melksham
Brigadier B.G. Wells
T. Wells
S. Wernham

J. White
T. White
Mrs. J. Wigmore, Chippenham
V. Wild, Marlborough
L. Williams
M. Wiltshire
Wiltshire Archaeological & Natural History Society
Wiltshire Ornithological Society
Wiltshire Wildlife Trust
Mr. & Mrs. M.G. Winfield-Chislett, Chippenham
Mrs. S. Wood, Bradford-on-Avon
Mrs. M. Woodhouse, Salisbury
Mr. D. Wyke, Poole
N.E. Wynn, Westbury
Miss J.M. Yates, Devizes
Miss J. Yates, Marlborough
Miss S. Yerrington, Bradford-on-Avon
Mr. M. Young, Salisbury
T. Young, London

Appendix 2
PUBLISHED WORKS BY OR ABOUT MARION BROWNE

This list includes all of Marion Browne's major publications. As she did not maintain a separate cuttings file for her own work, there is no definitive list and I may have missed some minor articles.

Mammal Reports:

Wilts. Nat. Hist. Mag.,	72 (1977), 30-39
	73 (1978), 32-40
	74 (1979), 38-41
	75 (1980), 34-40
WANHS Annual Report,	1981, 39-44
	1982, 57-61
	1983, xxii-xxvii
	1984, xxii-xxviii
	1985, xxvii-xxxiii
	1986, xxx-xxxvii
	1987, xxix-xxxv
WANHS Natural History County Recorders' Reports,	1988, 15

Systematic accounts:

Harvest mouse	*Wilts. Nat. Hist Mag.*,	70 (1975), 3-9 (with P. Dillon)
Water vole	*WAM,*	77 (1983), 123-137
Voles and mice		78 (1984), 94-104 (with P Dillon)
Dormouse		79 (1985), 206-213
Shrews		80 (1986), 197-210
Hedgehog		81 (1987), 111-122
Mole		82 (1988), 147-155
Common rat		83 (1989), 190-200

Other species accounts and themed articles:

Black specimens of *Vulpes vulpes. WAM,* 68 Part A (1973), 33-34.

Vixen *Vulpes vulpes.* Rearing nine cubs 1972. *WAM,* 68 Part A (1973), 34-35.

Vulpes vulpes. A Family on the move. *WAM,* 69 Part A (1974), 45-47.

The harvest mouse *(Micromys minutus)* in Wiltshire. *WAM,* 69 Part A (1974), 47-48 (with B. Gillam).

Vulpes vulpes. The Pig's Head, 1974. *Wilts. Nat. Hist Mag.*, 70 (1975), 32-34.

An orphan fox cub. *Wilts. Nat. Hist Mag.*, 71 (1976), 27-28.

The problem of Bovine TB in Badgers / Badgers and Wiltshire Naturalists. *WANHS Natural History Bulletin,* 6 (1976), 14-15.

The Water Vole. *WANHS Natural History Bulletin,* 7 (1977), 8-9.

Bat Bits. *WANHS Natural History Bulletin,* 7 (1977), 9-10.

The White Badger. *WANHS Natural History Bulletin,* 8 (1977), 11.

One Roe that ran. *WANHS Natural History Bulletin,* 10 (1978), 17.

The Hunted Fox. Summary of a paper read before a Joint Meeting of the Bristol Naturalists' Society and the Mammal Society, Bristol, March 11th, 1978. *WANHS Natural History Bulletin,* 10 (1978), 8.

Wildlife recording in the mechanical era. *WANHS Natural History Bulletin,* 12 (1979), 6.

Have you seen a Water Vole lately? *WTNC Newsletter,* September 1981.

Bats drive them potty in the loo. *Wiltshire Gazette and Herald,* August 8th, 1982.

Befriending Bats. *WTNC Newsletter,* September 1982.

Live trapping ... is it legal? *WTNC Annual Report,* 1993.

The Reptiles and Amphibians of Wiltshire Second Interim Report. *WANHS Annual Report,* 1982, 38-46 (with P. Dillon).

Know your small rodents. *BBONT Bulletin,* 74, September 1984, (with P. Dillon).

The Water Vole in Wiltshire. A Report on the 1981/82 Survey. *WTNC Newsletter,* September 1984.

Back to the Belfry – an injured pipistrelle restored. *WTNC Newsletter,* September 1985.

Bat Aid. First Aid and General Care of Bats. Wiltshire Bat Group, 1987.

Country Commentary in the *Wiltshire Gazette and Herald*:

Night a bat fell in grate. April 4th, 1981.

A moving tale about pig's head. June 18th, 1981.

Brush with a deadly bite. April 29th, 1982.

It's dreadful what people say about beautiful bats. June 10th, 1982

Hazards galore for a hedgehog. March 24th, 1983.

Be sure they are orphans. April 1983.

Furry friends deserve more interest. June 23rd, 1983.

Chance for those with a good nose for shrews. May 3rd, 1984.

Reports of meetings:

Spiders. Talk by Hugh Ennion, Devizes Museum, November 20th, 1975. *WANHS Natural History Bulletin,* 6 (1976), 6-7.

Porton Ranges, June 13th, 1976. *WANHS Natural History Bulletin,* 6 (1976), 9-11.

Great Cheverell Field Meeting, June 2nd, 1976. *WANHS Natural History Bulletin,* 7 (1977), 3.

Orchids of the British Isles. Talk by Ron and Joan Swanborough, Devizes Museum, October 7th, 1976. *WANHS Natural History Bulletin,* 7 (1977), 4.

Report on the Otter Conference, Andover, July 3rd, 1976. *WANHS Natural History Bulletin,* 7 (1977), 7-8.

Praise for Recorder of County's Moths, *Wiltshire Times and News,* February 18th, 1977.

Mammal Society at Bristol. *WANHS Natural History Bulletin,* 10 (1978), 14-15.

Deer and Forestry. Talk by Richard Prior at Devizes Museum, December 5th, 1977. *WANHS Natural History Bulletin,* 10 (1978), 3-4.

Fly High Run Free. Talk by Jane and Ted Ratcliffe at the AGM 7 May 1981 at Bromham School. *WTNC Newsletter,* September 1981.

Book Reviews:

Frank Sawyer and Sidney Vines. *Frank Sawyer, Man of the Riverside,* 1984. *WAM,* 79 (1984), 270.

Roy Pitman. *A Naturalist at Home,* 1984. *WAM,* 80 (1985), 259-60.

A.A. Wardhaugh. Bats of the British Isles. *WAM,* 83 (1990), 254.

Letters to Newspapers:

Wild animals have right to be left alone. *Wiltshire Gazette and Herald,* February 10th, 1983.

Countryside is not a free for all. *Wiltshire Gazette and Herald,* February 24th, 1983.

Countryside is not a free for all. *Wiltshire Gazette and Herald,* March 10th, 1983.

Keep up the vole roll. *Wiltshire Gazette and Herald,* March 17th, 1983.

Articles about Marion Browne and her work:

Bats quit belfry. *Swindon Evening Advertiser,* July 17th, 1978.

... And abandoned bats. *Wiltshire Times and News,* July 27th, 1978.

The Hunted Fox. *WANHS Natural History Bulletin,* 10 (1978), 4.

Marion Browne 1921-1990. *WTNC Magazine,* Autumn 1990.

Marion Browne. *WAM,* 84 (1991), 166-168.

Marion Browne. *The Mammal Society Newsletter,* 83, Autumn 1990.

Appendix 3
MARION BROWNE'S WORKS IN MANUSCRIPT

The following documents have been placed in Devizes Museum: all material dealing with species records in the Wiltshire Biological Records Centre; the rest in the Library.

1.0.0.0 **Natural History**

1.1.0.0 Contacts and correspondents in the Wiltshire Trust for Nature Conservation – card index, 5 x 3 boxed – name, address, telephone number, individual's identification code.

1.2.0.0 Contacts and correspondents in Wiltshire natural history societies – card index, 5 x 3, boxed – name, address, telephone number, individual's identification code.

1.3.0.0 Diaries 1971-89 – 24 volumes, various sized hardbound exercise books containing detailed field observations, including some information on weather, and diary entries.

1.4.0.0 Correspondence – extensive files of material mainly concerned with responsibilities and duties undertaken for the Wiltshire Trust for Nature Conservation and the Wiltshire Archaeological and Natural History Society, organised in the following sections:

1.4.1.0 Wiltshire Archaeological and Natural History Society (mainly Natural History Committee).

1.4.2.0 Wiltshire Archaeological and Natural History Magazine editorial matters.

1.4.3.0 Wiltshire Trust for Nature Conservation (mainly publicity and sales matters).

1.4.4.0 Lecturing.

2.0.0.0 **Mammals**

2.1.0.0 Small mammals:

2.1.1.1 Various species

2.1.1.2 Water vole

2.1.1.3 Harvest mouse

2.1.1.4 Dormouse

2.1.2.0 Carnivora:

2.1.2.1 Fox

2.1.2.2 Badger

2.1.2.3 Otter

2.1.3.0 Insectivora:

2.1.3.1 Hedgehog

2.1.3.2 Mole

2.1.3.3 Shrews

2.1.3.4 Bats

2.2.0.0 Species records:

2.2.1.0 Contacts and correspondents – card index, 5 x 3, boxed – name, address, individual's identification code (unique for each contributor and used in the acknowledgements section of each published paper and report).

2.2.2.0 Mammal record sheets 1975-84 (box file); 1984-89 (ring binder). Each contains record sheets from participants in the Wiltshire mammal survey, casual records and correspondence relating to both. Records (signified by a tick) subsequently entered into separate folders for each class of mammalia (see 2.2.5.0 below).

2.2.3.0 Mammal records – card index, 8 x 5 and 5 x 3, boxed – Wiltshire Biological Record Centre cards arranged by species – locality, grid reference, 10km sq., recorder, nature of record. Includes historical (pre-1960) cards with data abstracted from the literature by P. Dillon.

2.2.4.0 National Biological Records cards arranged by species – year, recorder's name and address, grid reference, vice-county, locality, nature of records.

2.2.5.0 Mammal files arranged according to class and species in A4 box files and ring binders:

2.2.5.1 Dormouse – distribution map, records map for 10km sq., date, grid reference, location, recorder, nature of record; correspondence, press cuttings.

2.2.5.2 Brown rat – as for dormouse.

2.2.5.3 Carnivora and Artiodactyla – maps and records only, by species.

2.2.5.4 Lagomorphs and squirrels – maps and records only.

2.2.5.5 Rodentia – maps and records only.

2.2.5.6 Insectivora – maps and records only.

2.2.6.0 Bats

2.2.6.1 Contacts and correspondents – card index, 5 x 3, boxed – name, address, telephone number, individual's identification code.

2.2.6.2 Records and distribution maps – A4 ring binder – date, vice-county, location, recorder and nature of record.

2.2.6.3 Card index (for records up to 1985) organised by species, mainly standard record cards with name and address of recorder, grid reference, vice-county, locality, date, nature of record.

From 1985 onwards arranged on an annual basis:

2.2.6.4 1985 record sheets, distribution maps, press cuttings and correspondence. (Note: some general mammal material in this file.)

2.2.6.5. 1986 – box file – detail as 2.2.6.4 above.

2.2.6.6 1987 – ring binder – records entered on Wiltshire Bat Group Questionnaire and NCC Bat Record Enquiry Form.

2.2.6.7. 1988 – box file – detail as 2.2.6.4 above.

2.2.6.8 1989 – ring binder – detail as 2.2.6.4 above.

2.2.6.9 Chiroptera Church Surveys for 1985-88 – box file.

3.0.0.0 **Reptiles and Amphibians**

3.1.0.0 Contributors and correspondents – card index, 5 x 3, boxed – name, address, telephone number, individual's identification code.

3.2.0.0 Correspondence, record sheets and contributors' raw data for Wiltshire Reptile and Amphibian Survey – A4 ring binder.

3.3.0.0 Records and distribution maps arranged by species and 10km sq. – habitat, date, grid reference, site, recorder, nature of record.

3.4.0.0 Proposed publication on reptiles and amphibians.

3.4.1.0 Correspondence and matters pertaining to proposed publication including minutes of meetings and draft contract – A4 ring binder.

3.4.2.0 Contributors' bibliography (compiled by P. Dillon) – author, date, title, place of publication, publisher – 5 x 3 index cards.

3.4.3.0 Fossil bibliography – "Wiltshire Fossil Vertebrata Part 2 Reptilia" (compiled by J.B. Delair) – author, date of publication, title, location.